Sensors 2020

Topical Conference at the 2020 AIChE Annual Meeting

Online
16 – 20 November 2020

ISBN: 978-1-7138-2292-9

Printed from e-media with permission by:

Curran Associates, Inc.
57 Morehouse Lane
Red Hook, NY 12571

CURRAN ASSOCIATES INC. proceedings.com

Some format issues inherent in the e-media version may also appear in this print version.

Printed with permission by Curran Associates, Inc. (2021)

For permission requests, please contact AIChE
at the address below.

AIChE
120 Wall Street, FL 23
New York, NY 10005-4020

Phone: (800) 242-4363
Fax: (203) 775-5177

www.aiche.org

Additional copies of this publication are available from:

Curran Associates, Inc.
57 Morehouse Lane
Red Hook, NY 12571 USA
Phone: 845-758-0400
Fax: 845-758-2633
Email: curran@proceedings.com
Web: www.proceedings.com

TABLE OF CONTENTS

611723 Investigating SERS of Electrogenic Bacteria Via Charge Transfer Enhancement By Graphene Quantum Dots (GQDs)

Abstract:

We report on the enhancement of the phononic excitation in electrogenic bacteria due to the electron transfer to interfaced graphene quantum dots (GQDs). This phenomenon enables GQDs to detect otherwise unobservable bacterial Raman peaks by chemeo-phononic amplification by a factor of ten. We show confirmation of the charge transfer ability of the electrogenic bacteria into GQD and provides a platform for chemical enhancement phenomena for biological sensing at the cellular level. This study also reports on the electrogenic bacteria's phonon-polarization directly related to the chemical enhancement from the GQDs.

608557 Near-Infrared Optical Detection of Dopamine Based on Xeno Nucleic Acid (XNA) Sensors

Abstract:

Single-walled carbon nanotubes (SWCNTs) emit near-infrared fluorescence that is beneficial for in vitro and in vivo optical sensing. State-of-the-art SWCNT sensors are functionalised with single-stranded DNA (ssDNA) sequences that control the dispersivity, brightness, biocompatibility, and selectivity of the sensor. However, the fluorescence of the ssDNA-functionalized nanotube sensors is sensitive to changes in the local ion concentration. This sensitivity is problematic not only for measurements in complex biofluids, but also for sensing neurotransmitters, whose concentrations are strongly coupled to spatial and temporal ion fluctuations. The fluctuating ion concentrations can therefore interfere with the accurate optical detection of neurotransmitters, thereby limiting the use of ssDNA-SWCNT sensors for detecting and treating neurological diseases.

We present an optical SWCNT dopamine sensor with improved stability towards cation-induced fluorescence changes. The synthetic biology approach presented in this work serves as a complementary means for enhancing nanotube optoelectronic behaviour, unlocking previously unexplored possibilities for developing nano-bioengineered sensors with augmented capabilities.

608015 A Wearable Microfiber Biomaterial Incorporates Optical Nanosensors and Enables Wireless Monitoring of Oxidative Stress

Abstract:

Reactive oxygen species (ROS) are a family of signaling molecules continuously generated, transformed, and consumed by all aerobic organisms. Wound healing is one of the most dynamic biological process featuring ROS-linked cellular signaling throughout the entire process. Additionally, basal concentrations of ROS aid in the fight against invading microorganisms into open wounds. The excessive and uncontrolled production of ROS contributes to the sustaining and deregulation of inflammation processes, which play a central role in the pathogenesis of chronic non-healing wounds. Physiologically, hydrogen peroxide (referred herein as peroxide) and superoxide serve as intracellular ROS messengers stimulating key phases of wound healing including cell recruitment, production of cytokines, and angiogenesis. Of note, peroxide acts as the principal secondary messenger in wound healing and is present at low concentrations (100–250 μM) in normal wounds. Increased peroxide concentration is a biomarker for chronic wounds in which biofilm-making pathogens can grow significantly faster than normal acute wounds. Due to the short half-life of peroxide, detection and quantification of their concentrations are difficult in vivo and in patients. Thus, peroxide levels in wounds have generally been determined indirectly through analysis of downstream oxidation products of lipids, proteins, or DNA.

Single-Walled Carbon Nanotubes (SWCNTs) with engineered wrappings have recently been developed and utilized in various disparate fields ranging from additives that strengthen material composites to biomedical applications including targeted drug delivery, optical biosensing, and biological imaging. Of significant interest, SWCNTs exhibit intrinsic photoluminescence (fluorescence) that is photostable, tunable, and sensitive to its local environment. The electronic band gap energies of SWCNTs are dependent on their chiral identity and are differentially and reversibly oxidized, enabling the ratiometric detection of local redox environments.

With revolutionary advances in nanotechnology and biomaterials in the recent years, an extensive range of smart wound care biomaterials have been developed which incorporate components such as growth factors and/or antimicrobial agents such as iodine or silver. Electrospun micro/nanofibers are one of the novel classes of wound dressings as they mimic the chemical and mechanical environment of the 3D extracellular matrix. Micro/nanofiber-based wound dressings have been designed to enhance cell migration, prevent inflammation and infection and inhibit scar formation on wounds. Herein, we developed a process to construct wearable textiles incorporating the DNA-wrapped SWCNT nanosensors through a one-step co-axial electrospinning method. We then developed an optical microfibrous biomaterial with a "core-shell" morphology that encapsulates fluorescent SWCNT nanosensors. We chose Polycaprolactone (PCL) as the shell material as it is an FDA-approved polymer which has been extensively studied for tissue engineering and wound healing applications. Our optical biomaterial was able to ratiometrically detect peroxide in the wound biological range (20-500 uM). The ratiometric characteristic of the sensor will significantly facilitate the future in vivo and clinical applications as it transduces an absolute signal which is not dependent on excitation source distance and exposure time. Moreover, the optical nature of the biomaterial will enable the wireless detection without a need to disturb the wound dressing. We revealed that the signal from the biomaterial does not significantly decay over time as the fluorescence of SWCNTs is non-photobleachable. We also quantified the absolute weight of SWCNT nanosenors in the biomaterial over time using confocal -Raman microscopy and indicated that the nanosensors do not leak out of the fibers over time, mitigating any toxicological concerns related to SWCNTs. Our optical biomaterial was integrated into other common wound dressing materials such as hydrogels and commercialized bandages, showing the compatibility of this platform with other existing wound dressings. Thus, our wearable microfiber biomaterial enables wireless monitoring of oxidative stress directly from the wound site.

610131 Inverse Emulsion Synthesis of Hydrogel-Coated Gold Nanoshells for Protein Biomarker Quantification

Abstract:

As proteomics research has begun to identify increasing numbers of differences in the protein makeup of patient samples, such as saliva, tears, urine, or serum, between healthy and diseased patients, there has been growing interest in developing new biosensor diagnostic tools that allow for minimally-invasive quantification of these potential biomarkers. One strategy is differential sensor arrays that exploit semi-selective recognitive elements in order to screen for differences in multiple protein biomarkers in a single assay. However, accurate protein recognition is challenged by the similarity of the target protein's structures and properties to those of the other proteins present in the sample, necessitating the use of multiple sensing elements in order to separate the interactions. In order address this challenge, we explore the capabilities of using inverse emulsion polymerization to form libraries of multifunctional hydrogel networks on the surface of gold nanoshells (AuNS) as semi-selective recognition elements for use in a localized surface plasmon resonance-based biosensor for the detection of autoimmune disease biomarkers.

Hydrogel nanoparticles were formed by an inverse emulsion copolymerization of methyacrylic acid with acrylamide or *N*-isopropylacrylamide, crosslinked with *N,N'*-methylenebis(acrylamide), and characterized by dynamic light scattering, potentiometric titration, and Fourier-transform infrared spectroscopy. AuNS were prepared using the method of seeded growth of gold colloids on aminated silica nanoparticles followed by encapsulation with poly(maleic anhydride-*alt*-1-octadecene)-g-poly(ethylene glycol) methacrylate (PMAO-g-PEGMA) graft copolymer to improve particle stability. Hydrogel-coated AuNS were prepared using the same nanoparticle synthesis method with the addition of PMAO-g-PEGMA AuNS to the aqueous phase of the inverse emulsion. UV-Vis spectroscopy was used to measure the shifts in the wavelength of the localized surface plasmon resonance for the hydrogel-coated AuNS. The successful synthesis of coated AuNS with both formulations shows immense promise for the creation of a library of hydrogels with various functional groups that could be used to alter affinity for different protein biomarkers from patient biofluids.

Work supported by National Institutes of Health Grant R01-EB022025

606325 Imaging Healthy and Diseased Striatal Dopamine Release with Near-Infrared Catecholamine Nanosensors

Abstract:

Neuromodulators such as dopamine and serotonin and play a critical role in healthy brain function, and their disruption has been implicated in a number of neurological and psychiatric disorders. In particular, the neuromodulator dopamine is thought to facilitate learning, motivation, and motor control. Disruption of dopamine signaling plays a crucial role in a number of neurological and psychiatric disorders, including Parkinson's Disease, Huntington's Disease, and Schizophrenia. Present treatments for these disorders often broadly target dopamine neuromodulation, and many novel therapies (RNAi, cell replacement) target circuits that participate or are affected by dopaminergic systems within the brain. However, despite the established relationship between these disorders and dopamine, the exact mechanisms through which these diseases progress, and their therapies act, is unknown. This is in part because our ability to study dopaminergic modulation in healthy and diseased tissue has been limited. Unlike classical neurotransmission, dopamine exerts it influence by diffusing beyond the synaptic cleft to influence the excitability of neighboring neurons. As such, being able to study the release and spread of dopamine at the spatio-temporal resolution of single dopamine release sites (μm, ms) is critical.

Here, we implement the near-infrared fluorescent catecholamine nanosensors (nIRCats) to image dopamine release within the brain striatum of R6/2 Huntington's Disease Model (R6/2) mice. We have previously shown that these near-infrared fluorescent, polymer-functionalized semiconducting single wall carbon nanotubes serve as adept dopamine sensors in the basal ganglia, exhibiting $\Delta F/F$ of up to 24-fold in the fluorescence emission window of 1000-1300 nm, a wavelength ideal for imaging in optically scattering brain tissue[1]. Furthermore, the chemically synthetic molecular recognition elements of nIRCats readily allows imaging of dopamine dynamics in diseased tissues or tissues that have undergone treatment via novel therapy without the need for additional genetic manipulation or mouse-line development. These unique properties of nIRCat make it an ideal tool to study the dysregulation of dopamine over the course of Huntington's Disease and subsequent treatment. We demonstrate that nIRCats imaging in R6/2 Mice *ex vivo* brain slices before and after the onset of motor symptoms reveal insights into possible mechanisms of dopamine decrease with Huntington's Disease motor symptom development. These findings provide a framework to understand how the diseased brain adapts to disease progression and subsequent treatment, and underscore the utility of nIRCats as a versatile new optical tool for dopamine detection in the brain.

1. Beyene, A. G., Delevich, K. et al. "Imaging Striatal Dopamine Release Using a Non-Genetically Encoded Near-Infrared Fluorescent Catecholamine Nanosensor." *Science Advances*, **5** *(7)* (2019).

600826 Varying Oleylamine to Dibenzyl Ether Ratio for Fine-Tuning Manganese Oxide Nanoparticle Diameter and MRI Signal Intensity

Abstract:

The ultimate goal of this research project is to develop a contrast agent to reduce the number of false positive and false negative imaging results associated with current breast cancer diagnostic imaging techniques. Gadolinium chelates are the current clinically approved contrast agents for breast magnetic resonance imaging (MRI), but they are always "on" and highlight any vascularized structure. Due to their lack of targeting, both benign and malignant breast tumors are enhanced with gadolinium. By utilizing tumor-targeted pH-sensitive manganese oxide (MnO) nanoparticles, a contrast agent can be developed that will only turn "on" in the presence of the lower pH of endosomes/lysosomes inside cancer cells. Reducing the size of MnO nanoparticles should increase dissolution of MnO to Mn^{2+} at low pH to enhance MRI signal. MnO nanoparticles were synthesized through thermal decomposition of Mn(II) acetylacetonate in dibenzyl ether (DE) and oleylamine (OA) at a peak reaction temperature of 280°C for 30 minutes. The following ratios of OA:DE were used: 60:0, 50:10, 40:20, 30:30, 20:40, and 10:50 to investigate the effects of reactant conditions on nanoparticle size, Mn^{2+} release and MRI signal. MnO nanoparticles were characterized physically and chemically using transmission electron microscopy (TEM), x-ray diffraction (XRD), and Fourier-transform infrared spectroscopy (FTIR). According to TEM, the 10:50 ratio created the smallest nanoparticle diameter of 19 ± 6 nm, while the 50:10 ratio created the largest nanoparticle diameter of 39 ± 15 nm. All nanoparticles were comprised of MnO (XRD) coated with oleylamine (FTIR). To further elucidate the effect of nanoparticle size and composition on the release of Mn^{2+} and resulting MRI signal, the smallest MnO nanoparticles (19 nm) were compared with MnO nanoparticles of a larger diameter (32 nm) and Mn_3O_4 nanoparticles of similar diameter (17 nm). Nanoparticles were incubated at 3 different pHs to measure Mn^{2+} release over time including pH 7.4 (blood pH), 6.5 (pH of tumor extracellular space), and 5 (endosome/lysosome pH) using inductively coupled plasma-optical emission spectroscopy (ICP-OES) and 1T MRI. The smallest MnO nanoparticles were the most efficient MRI contrast agents by releasing the highest amount of Mn^{2+} at pH 5 (~50% at 24 hr) and 6.5 (~11% at 24 hr), which produced the lowest T_1 value (646 ms at 24 hr, pH 5) and maximum MRI signal at 1T; Mn_3O_4 nanoparticles generated the lowest MRI signal. Therefore, MnO nanoparticles instead of Mn_3O_4 should be utilized, and the nanoparticles should be targeted intracellularly to maximize MRI contrast generation. Future work includes fabricating hydrophilic MnO nanoparticles using phospholipid encapsulation and adding breast cancer-targeting peptides to enable specific detection of malignant versus benign breast tumors.

597260 Understanding Corona Exchange Dynamics on Carbon Nanotubes with Multiplexed Fluorescence Monitoring

Abstract:

Adsorption of polymers on single-walled carbon nanotubes (SWCNTs) has enabled developments in molecular sensing, *in vivo* imaging, and gene delivery applications.[1,2] Noncovalent functionalization of SWCNTs with polymers preserves the pristine SWCNT atomic structure, thus retaining the intrinsic near-infrared fluorescence for sensing or imaging functions, and offers a reversible binding mode for cargo delivery. However, noncovalent biomolecule adsorption is an inherently dynamic process, where exchange occurs between molecules in the bulk solution and molecules on the nanoparticle surface, into what is known as the corona phase. The nature, strength, and kinetics of both the polymer binding and unbinding processes on SWCNTs are important contributors to the success of polymer-SWCNT based technologies. Understanding this binding process is especially important for intended uses of functionalized SWCNTs in biological environments, where native biomolecules compete with the original polymer to occupy the SWCNT surface. Binding of proteins and other biomolecules to the SWCNT disrupts the intended functionality of the nanoparticle and can further lead to adverse biocompatibility outcomes.

We present an assay to study the corona exchange dynamics between solution-phase and corona-phase polymers on SWCNTs.[3] This assay exploits the quenching property of fluorophores proximal to the SWCNT surface to monitor ligand binding and unbinding events. Real-time tracking of polymer adsorption and desorption events on the SWCNT surface are conducted with systematic variation of the molecular entities and solution conditions, with a specific focus on DNA (as the sensing moiety) and protein (as the biofouling agent). Binding profiles are extracted from the experimental assay and used to inform a kinetic model of the system. The work presented herein develops an understanding of the fundamental corona exchange mechanism and provides insight into performance of these designed SWCNT-based systems in biologically relevant, protein-rich conditions.

References

1. Demirer, G. S. *et al.* High aspect ratio nanomaterials enable delivery of functional genetic material without DNA integration in mature plants. *Nature Nanotechnology* 1 (2019) doi:10.1038/s41565-019-0382-5.
2. Beyene, A. G. *et al.* Imaging striatal dopamine release using a nongenetically encoded near infrared fluorescent catecholamine nanosensor. *Science Advances* **5**, eaaw3108 (2019).
3. Pinals, R. L., Yang, D., Lui, A., Cao, W. & Landry, M. P. Corona Exchange Dynamics on Carbon Nanotubes by Multiplexed Fluorescence Monitoring. *J. Am. Chem. Soc.* (2019) doi:10.1021/jacs.9b09617.

598813 DNA-Caged Polymer Nanocomposites for Erasable Fluorescence Imaging

Abstract:

Imaging pathology samples is often difficult because of limited colorimetric dye options and tissue autofluorescence resulting from structural complexity. If fluorescent labels are used, they are limited to color channels with minimal overlap with tissue autofluorescence, which is not ideal for image multiplexing. Sequential erasable labeling with fluorescent antibodies has been proposed as a potential solution, often with the use of harsh chemicals to remove the antibodies. These chemicals can damage tissue, which may lead to loss of structural information or misidentification of important structures. Newer processes featuring single stranded DNA (ssDNA) modified with fluorophores as reporters may address these limitations. DNA labels can be easily erased by strand displacement reactions. However, unprotected fluorophores are easily bleached and lack stability. Here, we report the creation of DNA caged-micelles (i.e., DNA cages) for use as an erasable fluorescent labeling system. DNA cages are DNA nanostructures (modified from Kurokawa et al.[1]) that form a network on spherical surfaces (i.e., micelles). The proposed labeling system is based on gentle DNA dehybridization interactions and labels are protected by micelle coatings, which would greatly improve current technology. Additionally, the dyes used in cell and tissue labeling can include colorimetric, fluorescent, and quantum dot labels.

Proof of cage formation was measured using DNA tiles modified with fluorescent dyes and/or fluorescence quencher FRET pairs, as well as via TEM images. Three different fluorescent curves were generated based on interaction of tiles with the micelle surface and/or each other. Absorption of quenching DNA tiles to a fluorescent micelle surface indicated binding of DNA cages to the micelle surface, with complete quenching at a molar ratio of 13.5:1 (DNA:PEG (micelle)). This result was validated by TEM images showing formation of a new structure not observed in the absence of DNA or polymer (micelles). Additionally, cage saturation on the micelle surface was observed at a molar ratio of 18:1 and interlocking of the DNA tiles (based on a FRET quencher pair) was observed at a molar ratio of 9:1.

Erasable labeling was demonstrated in solution by binding the targeting strand on DNA cages to ssDNA labeled with fluorophores that was then erased in sequential cycles via strand displacement reactions. In solution, an erase depth of greater than 90% was achieved for multiple cycles. DNA cages were then loaded with fluorescent dyes and used as labels in fixed cells. U87 glioblastoma cells were labeled with primary antibodies targeting either integrin β1 (an extracellular target) or actin (an intracellular target). Then, secondary antibodies conjugated to ssDNA were used to bind primary antibodies. Finally, cages containing complementary ssDNA to that on the secondary antibodies were introduced for target labeling. Labeling of both targets was observed, consistent to that of gold standard controls. Next, DNA with greater complementarity to secondary antibodies was introduced, erasing DNA cages. An erase depth of up to 85% was achieved. DNA cages could also penetrate up to 100 μm in tissue sections with no damage observed over the course of treatment and washing. These results suggest that DNA cages show great promise for use in multiplexed cell labeling.

Reference

1 Kurokawa, C. *et al.* DNA cytoskeleton for stabilizing artificial cells. *Proc Natl Acad Sci U S A* **114**, 7228-7233, doi:10.1073/pnas.1702208114 (2017).

600369 Development of an Electroactive Platform for Detection of Virus Fusion to Host Membranes

Abstract:

Every year, new virus outbreaks are reported and have potential to cause pandemics with devastating socio-economic impacts, with the coronavirus disease 19 (COVID-19) serving as a prime example. Recent viral outbreaks in this decade have alluded to the importance of quick and accurate viral sensing strategies so that the infected individual can begin their treatment and limit contact with uninfected individuals quicker, mitigating the impact of an outbreak. PCR, one of the most common viral detection methods, is a specific and sensitive detection technique, but requires trained personnel and sophisticated machinery, which can slow response times. An ideal detection method should integrate the specificity and sensitivity with the speed necessary to diagnosis patients rapidly.

Enveloped viruses (such as influenza virus, coronavirus) enter host cells by binding to a receptor and then fusing its viral membrane to the host cell membrane to insert its genome. We propose a novel sensing platform that leverages the virus's propensity to bind specifically and fuse to a host cell membrane. We can couple a biomimetic membrane on top of an electrically conductive polymer (PEDOT:PSS) to probe the electrical responses of the system once viral fusion has occurred. Successful viral fusion to the membrane in a matter of minutes may cause a detectable change in electrical response, serving as the readout of a viral infection. Using influenza virus as a proof-of-principle, we first show that we can form the biomimetic membrane incorporating the influenza viral receptor on PEDOT:PSS. We then visualize influenza single virion fusion on PEDOT:PSS and demonstrate that the kinetics of this fusion event are comparable to influenza fusion on a non-conductive polymer. Lastly, using electrical impedance spectroscopy, we provide evidence that influenza viral fusion causes a detectable change in electrical response, demonstrating that it is possible to exploit viral membrane fusion as a diagnostic strategy.

599943 Engineering DNA-Based Materials for the Analysis of Live Single Cells

Abstract:

The ability to probe the composition of live cells is important for learning about chemical processes inside of cells as well as for creating diagnostic platforms based on molecular profiling. However, current strategies for detecting analytes in single living cells are severely limited both in terms of probe design and sensing strategy. While fluorescent proteins and nucleic acid-based aptamers have revolutionized live-cell imaging, these sensors must be genetically encoded into the cells and therefore have limited applications in a clinical setting. Many techniques require cell lysis or a bulk population of cells to yield a detectable quantity of the target analyte. Additionally, a major impediment to the development of intracellular probes is the inability of most probes to traverse the cell membrane. Consequently, these probes require the use of transfection reagents or physical methods that perturb the cell membrane, which in turn can interfere with the measurements by changing the analyte's expression profiles. Furthermore, existing strategies for signal transduction to indicate the presence of a given analyte, such as fluorophore-quencher-based systems and FRET-based methodologies, are limited and susceptible to false-positive signal. Therefore, the development of a versatile platform for intracellular analyte detection in live cells is an outstanding challenge.

To address this challenge, I have first developed DNA-based aptamers that utilize a novel sensing strategy. Specifically, these aptamers are chemically modified at a single base with a viscosity-sensitive dye. Target binding to the aptamer changes the conformation of the aptamer and forces the dye to intercalate (or FIT) between the DNA base pairs, turning its fluorescence on. This "FIT" strategy avoids false-positive signals common to most fluorophore-quencher-based systems, quantitatively outperforms FRET-based techniques, gives up to 20-fold fluorescence enhancement in buffer, and allows for the detection of analytes in complex biological milieu such as human serum.

To apply this technique to the analysis of intracellular analytes, I have then developed a new platform based on protein spherical nucleic acids (ProSNAs). ProSNAs are composed of a protein core surrounded by a DNA shell. The dense functionalization and radial arrangement of the DNA around the protein core allows efficient cellular uptake of the ProSNA without the use of transfection reagents or physical stimuli. By using a FIT-aptamer as the DNA sequence, this design allows for the detection of intracellular analytes. As an example, I demonstrate that using an i-motif sequence that undergoes pH-dependent conformational changes allows for the detection of intracellular pH. As more than 500 aptamers are known for various targets, our nano-construct is a powerful platform for the detection of intracellular analytes. Moreover, by using a functional protein as the core, it is possible to detect analytes for which aptamers do not exist and further expand the range of analytes that can be detected. Using glucose oxidase as the core, I show that glucose can be detected (for which is there is no known aptamer with a biologically relevant binding affinity) in 8 different cell lines. Taken together, these constructs represent a plug-and-play platform for detecting a wide variety of intracellular analytes with great potential in the field of biodiagnostics.

References

†Equal Contribution

Samanta, D.†; **Ebrahimi, S. B.†**; Mirkin, C. A., "*Nucleic-Acid Structures as Intracellular Probes for Live Cells,*" *Advanced Materials* **2020**, *32*, 1901743.

Ebrahimi, S. B.†; Samanta, D.†; Cheng, H. F.; Nathan, L. I.; Mirkin, C. A., "*Forced Intercalation (FIT)-Aptamers,*" *Journal of the American Chemical Society* **2019**, *141*, 13744.

Samanta, D.; Iscen, A.; Laramy, C. R.; **Ebrahimi, S. B.**; Bujold, K. E.; Schatz, G. C.; Mirkin, C. A., "*Multivalent Cation-Induced Actuation of DNA-Mediated Colloidal Superlattices,*" *Journal of the American Chemical Society* **2019**, *141*, 19973.

608845 Molecular Engineering of a Colorimetric Nanogel Sensor for Clinical Radiotherapy and Trauma Monitoring

Abstract:

Radiotherapy using ionizing radiation (e.g. X rays) remains a mainstay of treatment modalities for different types of cancers. Overexposure to radiation in the clinic or as part of an adverse event can induce tissue toxicity and damage, whereas underdosing can lead to poor efficacies of tumor ablation. Effective radiation sensors are critical in ensuring optimum delivery to patients undergoing radiotherapy and for patient stratification in cases of radiation exposure. Currently existing sensors possess several disadvantages in rapid radiation detection because of their fragility, sensitivity to light and heat, dose-rate dependence, long processing times, lack of conformation to tissue morphologies and / or high costs. Commonly used Nanodots™ are expensive, require separate read-out devices and do not conform to tissue morphologies. We developed a novel colorimetric nanogel sensor, based on the formation of maroon-colored gold nanoparticles from their colorless precursor salt formulations, upon exposure to different levels and types of ionizing radiation. By measuring the intensity of color developed with a simple bench-top absorbance spectrophotometer we were able to detect doses typically administered in fractionated cancer radiotherapy (i.e. ranging from 0-5 Gy doses) and beyond. Topographical distribution of delivered radiation doses was qualitatively and quantitatively determined following addition of quenching agents, and the efficacy of the nanogel sensor was demonstrated for different sources of ionizing radiation including protons, electrons, photons and radioactive isotopes. The translational potential of the sensor was demonstrated using anthropomorphic phantoms and in live canine patients undergoing clinical radiotherapy treatments. Our results demonstrate a new generation of nanogel sensors for colorimetric detection of radiation for improving patient safety in clinical radiotherapy and trauma care.

609804 Unique Electrochemical Detection of Sepsis Using Triplex Biomarker Detection Panel with IL-6, IL-8 and IL-10 in Blood Plasma

Abstract:

Sepsis is a life-threatening disease that is caused by the body's extreme reaction to an infection which could lead to tissue damage, multiple organ failure or even death. It is one of the leading causes of death worldwide, estimated to affect more than 30 million people every year according to the world health organization. Despite scientific developments in health care technologies in the past three decades, a substantial challenge continues to persist for early rapid diagnosis of sepsis. Thorough and timely diagnosis plays a substantial role in increasing the survival chances for patients with adequate treatment, thus reducing mortality. However, current gold standard blood culture techniques require large patient sample volumes and provides delayed results (6 hours to up to 5 days). This may lead to clinical ambiguity in treatment measures, propelling the body's immune response to spiral uncontrollably into septic shock or severe sepsis. This fails to deliver the required speed and accuracy to enable accurate diagnosis. To overcome existing challenges, we tailored a specific biosensor to screen for sepsis as a rapid point-of-care (POC) technology enabling multiplexed detection capabilities. The novelty presented in this research combines the effectiveness of specific multiplexed biomarkers for detection of sepsis with electrochemical impedance spectroscopy (EIS) technique to enable clinically efficient decision-making process. This work leverages the detection of pro and anti-inflammatory cytokine biomarkers such as interleukin 6 (IL-6), interleukin-8 (IL-8) and interleukin-10 (IL-10) in blood plasma. Researchers suggests good correlation between timing of pro and inflammatory cytokine release to degree of sepsis, associated with deleterious effects. The developed label-free, point-of-care biosensor is surface engineered with specific capture probes that utilizes EIS to measure the capacitive impedance change reflecting binding interactions between the capture probe and target biomarker enabling simultaneous triplex detection. Specificity of the POC biosensor was validated using cross-reactive studies, which displayed insignificant interference from non-specific biomarkers. The biosensor also displays stable and repeatable performance. Key advantage of the triplex biosensor is the use of undiluted blood plasma volume of less than 40 μL, providing result within 15 minutes. Additionally, the triplex biosensor technology was validated with 20 patient samples that were clinically tested positive for sepsis. This unique sensor technology would allow medical practitioners to facilitate targeted interventions for septic patients as a rapid prognostic approach, preventing complications arriving from sepsis.

598873 Harnessing the Protein Corona Towards Carbon Nanotube-Based Sensor Design

Abstract:

Unpredictable protein adsorption on nanoparticles remains a considerable challenge towards effectively applying nanotechnologies, such as nanosensors, in biological environments. However, if well understood and controlled, proteins adsorbed on nanoparticles can be exploited to develop protein-based nanosensors.

Herein, we employ our knowledge of spontaneous "protein corona" formation on single-walled carbon nanotubes (SWCNTs) in biofluids towards rational nanosensor design based on protein-SWCNT complexation.[1] SWCNTs are implemented as the sensor platform due to their desirable optical and physical characteristics, including photostable fluorescence in the near-infrared tissue transparency window[2] and high aspect ratio that confers delivery capabilities.[3] We first study the protein corona composition formed on SWCNT-based probes within relevant biofluids such as blood plasma and cerebrospinal fluid.[4] Then, we take advantage of the intrinsic binding affinity of specific proteins on the carbon nanotube to selectively adsorb proteins that, in turn, bind to their natural ligand partners. Thus, the adsorbed protein acts as the recognition moiety that interacts with the environment and detects the analyte of interest, tethered to the SWCNT that serves as the signal transducer. This nanosensor platform combines the sensitive and selective binding capabilities of proteins for their ligand partners with the near-infrared fluorescence readout of SWCNTs that is modulated upon protein-ligand binding. In this manner, we demonstrate that although formation of the nanoparticle protein corona can impede the intended nanobiotechnology application, it also presents an opportunity to create improved protein-nanoparticle architectures by manipulating selective protein adsorption on the nanoparticle surface.

References

1. Pinals, R. L., Yang, D., Lui, A., Cao, W. & Landry, M. P. Corona Exchange Dynamics on Carbon Nanotubes by Multiplexed Fluorescence Monitoring. *J. Am. Chem. Soc.* (2019) doi:10.1021/jacs.9b09617.
2. Boghossian, A. A. *et al.* Near-Infrared Fluorescent Sensors based on Single-Walled Carbon Nanotubes for Life Sciences Applications. *ChemSusChem* **4**, 848–863 (2011).
3. Demirer, G. S. *et al.* High aspect ratio nanomaterials enable delivery of functional genetic material without DNA integration in mature plants. *Nat. Nanotechnol.* **14**, 456–464 (2019).
4. Pinals, R. L. *et al.* Protein Corona Composition and Dynamics on Carbon Nanotubes in Blood Plasma and Cerebrospinal Fluid. *bioRxiv* 2020.01.13.905356 (2020) doi:10.1101/2020.01.13.905356.

607484 Point-of-Care Cancer Biomarker Detection System Integrating Surface Acoustic Wave Streaming and Metal-Enhanced Fluorescence

Abstract:

Point-of-care cancer biomarker detection system integrating surface acoustic wave streaming and metal-enhanced fluorescence

Yuqi Huang, Shuangming Li and Venkat R. Bhethanabotla

yuqi@usf.edu

shuangming@mail.usf.edu

bhethana@usf.edu

Department of Chemical and Biomedical Engineering, University of South Florida, Tampa, Florida 33620-5350, USA

Abstract:

Immunofluorescence is one of the primary techniques for microbiological assays. It can help to visualize, identify, and even quantify the interaction between antibody and antigen, some even used for DNA and protein detection.[1,2] However, there is a clinical need of increasing its limit of detection, specificity and portability. In this work, a portable optical fluorescence system was designed to be combined with a direct digital synthesizer for surface acoustic wave (SAW) generation and metal-enhanced fluorescence (MEF) to lower the detection limit in immunofluorescence assays.[3,4] MEF was achieved by depositing silver nanoparticles with covered silica film, which can plasmonically enhance fluorescence intensity. Rayleigh wave streaming was generated on a piezoelectric device for surface acoustic wave streaming purpose, which can prevent nonspecific binding on the biological interface.[5] The miniaturized prototype, integrating acoustic wave removal of interfering proteins combined with optical detection, was developed to realize immune detection on a portable platform with limit of detection at nanogram per milliliter level in detecting carcinoembryonic antigen (CEA) from human plasma.

Recent work shows that the MEF substrate fabricated with the above method can lower the detection limit of CEA from PBS to 100pg/mL level. However, testing the detection limit for CEA from human plasma has raised to 500ng/mL. Non-specific proteins, dusts and electrolytes in human plasma are primary suspects to raise background noise which increases the limit of detection. Further tests show that by adding SAW streaming to the sample tested on MEF substrate, the detection limit brings back down to 50pg/mL. A reasonable inference to this detection limit drop is the removal of non-specific protein by streaming surface acoustic wave to the sample surface, where the non-specific absorbents are removed by shear force from the wave, and the background noise is reduced to lower the limit of detection for CEA from human plasma.

1. K. Aslan, I. Gryczynski, J. Malicka, E. Matveeva, J. R. Lakowicz, and C. D. Geddes, "Metal-enhanced fluorescence: an emerging tool in biotechnology," Current Opinion in Biotechnology, vol. 16, pp. 55-62, **2005**.
2. R. A. Goldsby, T. J. Kindt, B. A. Osborne, and J. Kuby, "Immunology," ed: New York: WH Freeman, **2003**.
3. S. Li and V. Bhethanabotla, “Design of a Portable Orthogonal Surface Acoustic Wave Sensor System for Simultaneous Sensing, Removal of Nonspecifically bound Proteins and Mixing”, Sensors, vol. 19(18), **2019**.
4. Liu, J.; Li, S.; Bhethanabotla, V. R., Integrating Metal-Enhanced Fluorescence and Surface Acoustic Waves for Sensitive and Rapid Quantification of Cancer Biomarkers from Real Matrices. *ACS Sensors*, *3* (1), 222-229, **2018**.
5. S. Cular, D. W. Branch, V. R. Bhethanabotla, G. D. Meyer, and H. G. Craighead, "Removal of nonspecifically bound proteins on microarrays using surface acoustic waves," IEEE Sensors journal, vol. 8, pp. 314-320, **2008**.

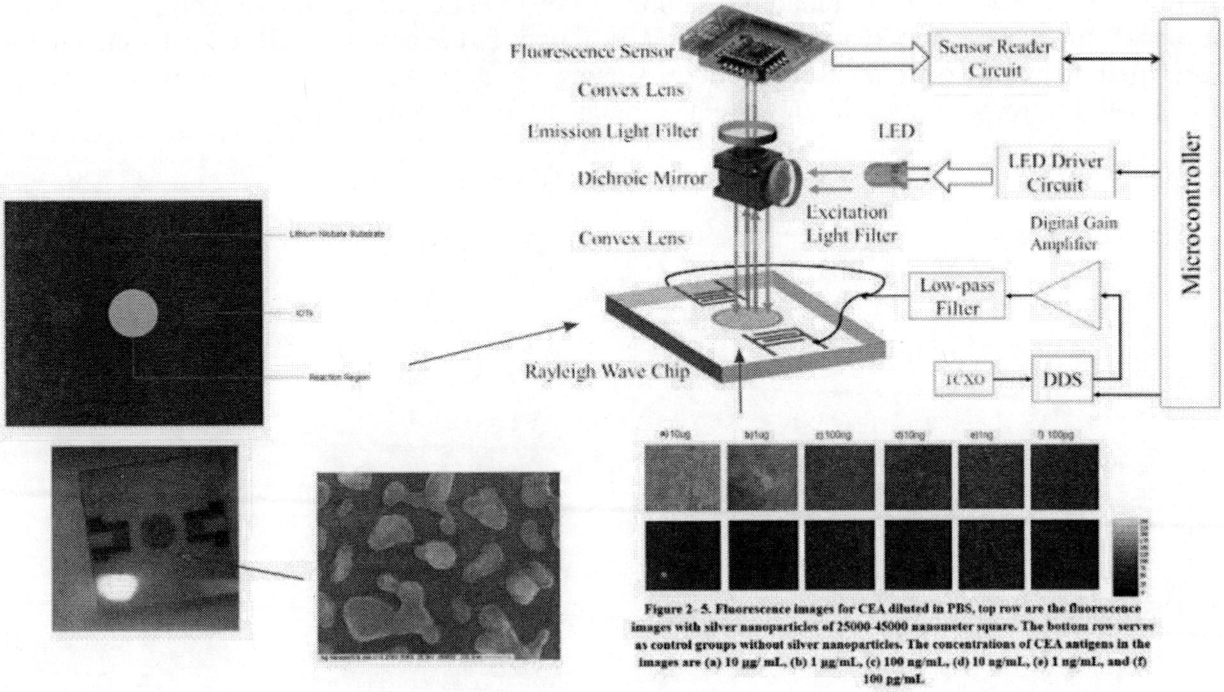

Figure 2-5. Fluorescence images for CEA diluted in PBS, top row are the fluorescence images with silver nanoparticles of 25000-45000 nanometer square. The bottom row serves as control groups without silver nanoparticles. The concentrations of CEA antigens in the images are (a) 10 μg/ mL, (b) 1 μg/mL, (c) 100 ng/mL, (d) 10 ng/mL, (e) 1 ng/mL, and (f) 100 pg/mL.

609302 High-Throughput Quantification of Influenza A Virus RNA Using Novel Drop-Based qRT-PCR Analysis

Abstract:

Influenza A virus (IAV) causes seasonal epidemics and sporadic worldwide pandemics that have resulted in over 100 million deaths during the past century. IAV populations are dynamic with high mutation rates and short replication times. A better understanding of IAV diversity at the single cell level would provide greater insight into how heterogeneity of infection affects population-level dynamics. Investigating IAV and host cell dynamics is typically performed using bulk culture and may not fully capture cell-to-cell differences in virus production. Thus, there is a need for high-throughput quantification of the IAV output, or burst size, from single cells to elucidate individual host-cell dynamics at the population scale. Here, we have developed a novel drop-based quantitative reverse transcriptase polymerase chain reaction (qRT-PCR) method to measure the amount of IAV RNA contained in individual drops. Concentrations of RNA are encapsulated in tens of thousands of picoliter-sized drops. These drops are processed on a microfluidic device which splits a small portion of the drop and merges it with qRT-PCR solution. The merged drops are thermocycled using a standard qPCR machine. Drops are randomly sampled at different cycles to measure drop fluorescence. Using a novel qPCR analysis method, we quantify the amount of RNA contained in the drops. Our method is the first to measure IAV RNA using a high-throughput drop-based assay, which will allow for future work in investigating the heterogeneity of IAV infection.

597969 Microfluidic Paper-Based Analytical Devices Using Plasma Processes

Abstract:

Disability adjusted life years (DALY) are the sum of years of life lost due to premature mortality and years lived with a disability. In 2010, DALY were approximately 2.4 billion. It is estimated that ~90% of DALY are due to communicable and non-communicable diseases. This problem is even more severe in developing countries where the major stumbling block for disease control is lack of proper disease diagnostic methods. Point-of care testing (POTC) methods in which users do not require specialized equipment and sophisticated lab facilities for disease diagnostics will be ideal for developing countries. Therefore, the World Health Organization (WHO) has stipulated a set of criteria for POCT in resource-limited settings called ASSURED criteria—affordability, sensitivity, specificity, user-friendly, rapid and robust, equipment-free and deliverable to those who need it. Due to the necessity and increasing demand of personalized medicine, the global POTC market is expected to grow from US $23.16 billion in 2016 to US $36.96 billion in 2021 at the compound annual growth rate of 9.8% from 2016 to 2021.

Microfluidic paper-based analytical devices (μ-PAD) have shown considerable promise to meet the ASSURED criteria set by the WHO. Paper has been patterned using various fabrication methods to allow distribution of liquid samples to targeted locations where subsequent reactions with reagents deposited in the paper are performed. These reactions develop color signals that can then be read using a smartphone. Low cost, portability, and the absence of external power source requirements make these devices desirable for POTC in less developed areas. It has been more than a decade since the first μ-PAD was invented. Yet these μ-PADs are not yet commercialized because of multiple technical problems including 1) contamination and evaporation of samples because of the lack of proper μ-PAD packaging, 2) inability to carry out complex tests that require multiple reagent additions, and 3) loss of sample in flow channels.

To address the issue of contamination and evaporation, a fully enclosed paper microfluidic device has been fabricated using pentafluoroethane (PFE) plasma deposition followed by O_2 plasma etching. Structures with one and two non-interacting, fully enclosed hydrophilic channels were generated in a single paper sheet using metal masks. Furthermore, by performing an additional O_2 plasma step with a secondary mask, pinholes were created at the reaction zones for reagent loading. Finally, to demonstrate the functionality of the device, a glucose assay was performed. Quantitative analysis of glucose assays showed that the device can be used for clinically relevant range of glucose. To our knowledge, this is the first time that such structures have been fabricated without paper stacking. Multi-layer devices have enhanced functionality relative to a single channel device, because the design space for creating networks of channels within the paper substrate is greatly expanded. The fluid-filled channels in the fabricated device are isolated, thereby preventing contamination due to handling and environmental exposure. Fluid evaporation can be inhibited by sealing the device with adhesive tape without contaminating the enclosed channels hence preventing loss of sample due to evaporation; a critical factor while working with precious samples.

To address the challenge of fluid flow control in μ-PADs, we extended our work on fully enclosed channel fabrication inside paper substrates to incorporate flow control functionality into the channels. Using pentafluoroethane plasma deposition and subsequent O_2 plasma etching with metal masks, fully enclosed hydrophilic channels with varying wettability were designed and fabricated to precisely control fluid flow in μ-PADs. Flow control is one of the most critical functionalities in μ-PADs because it allows controlled implementation of multiple sequential reactions on a single device with minimal user interference. Potential benefits of flow control include improved limit of detection, ability to carry out complex diagnostic tests, reduced assay time, and decreased likelihood of human error. We studied the effect of controlled fluorocarbon removal on the surface of cellulose fibers on channel wettability and demonstrated that this behavior can be exploited to significantly vary the fluid velocity in the flow channels. Furthermore, by simply covering the flow channels with adhesive tape, we demonstrated that the effect of evaporation on the fluid flow can be significantly decreased. Finally, a device with three channels converging to a single reaction zone was fabricated and sequential delivery of fluid to the reaction zone was demonstrated. The devices made by the methods described in this paper offer significant benefits for in-field diagnostic testing where multistep protocols are required.

A device fabricated by this method is estimated to cost between 10 and 50 ¢ and the plasma processes used to fabricate these devices are compatible with roll-to-roll technology, making it suitable for large scale production. This alternative fabrication approach contributes significantly toward commercialization of low-cost μ-PADs that meet the WHO ASSURED criteria.

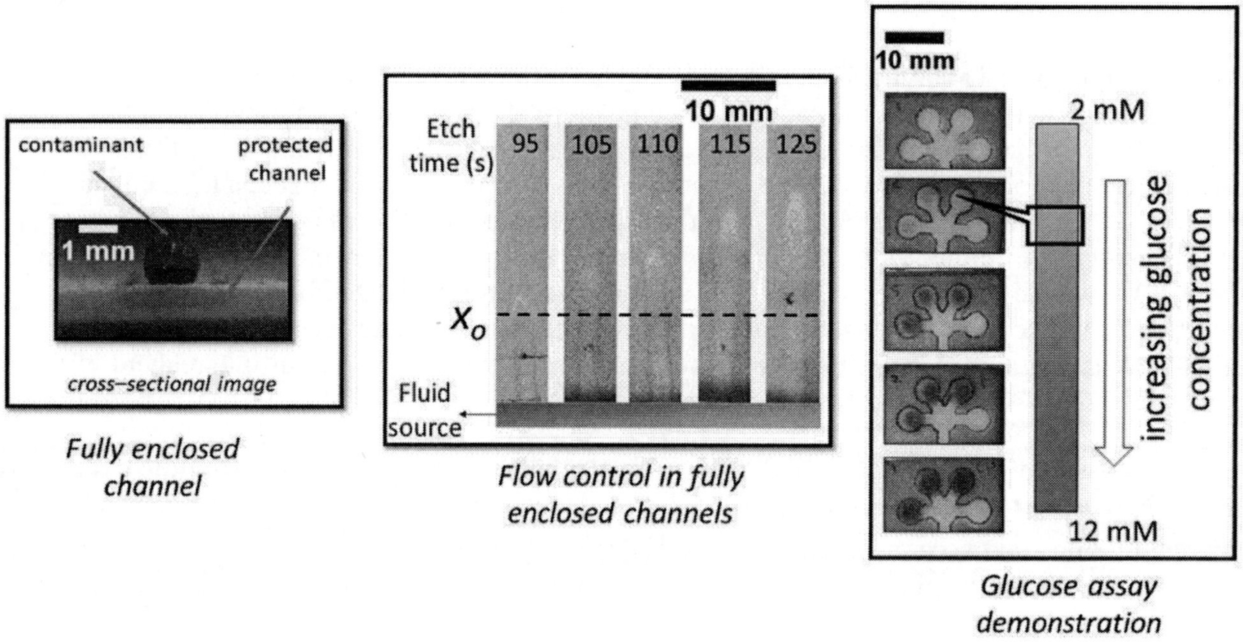

607198 Detection of Atrazine and Its Metabolites Using Photonic Molecularly Imprinted Polymers

Abstract:

Wetlands are beneficial in numerous ways and so, keeping them healthy and clean is an essential goal to fulfill. Due to use of herbicides for farming purposes, runoffs after raining would bring uncontrolled amounts of these herbicides to the wetlands. Atrazine is one of the most common herbicides used by farmers in the United States due to their selectivity against broadleaf weeds and annual grasses. Extensive use of this herbicide has caused contaminations in drinking water and soil and it can be harmful to living organisms even at low concentrations. According to EPA regulations, the drinking water maximum contaminant level (MCL) for atrazine is 3 ppb and being exposed to atrazine above this level may lead to lung, heart, and kidney diseases, low blood pressure, muscle spasms, weight loss, and damage to adrenal glands. Hence, it is important to monitor the level of atrazine in drinking water and ground water. Atrazine is degraded by plants and microorganism to metabolites which some of them are hazardous and detecting desethylatrazine (DEA) and desisopropylatrazine (DIA) is as essential as atrazine. The methods currently used to detect atrazine in different samples include gas chromatography (GC), gas chromatography-mass spectrophotometry (GC/MS), high pressure liquid chromatography (HPLC), liquid chromatography mass spectrophotometry (LC/MS) and enzyme-linked immunosorbent assay (ELISA). The gold standard for atrazine analysis, LC-MS, is precise but expensive, complex and time-consuming. The ELISA method, although relatively precise, has been found to be inaccurate in many cases and have positively biased results due to other organic materials interferences. Therefore, a method which is inexpensive, rapid and facile applicable is highly needed.

We developed an inexpensive label-free photonic sensor for the rapid detection of atrazine and its metabolites for field use. The sensor is composed of a molecularly imprinted polymeric (MIP) porous film that selectively captures atrazine molecules; the film pore structure is an inverse opal and its optical properties change when atrazine is captured.

Molecularly Imprinted Polymers (MIPs) have highly specific adsorption sites in their polymeric structure which leads them to selectively adsorb target molecules in those sites. Using a colloidal crystal of silica nanoparticles, a porous structure has been created in the polymer to facilitate target molecules movements through the thin film of polymer. By combining colloidal-crystal templating and molecular imprinting technique, a *p*H-paper-like sensor for a label-free colorimetric detection of atrazine and its metabolites in aqueous solutions and wetlands is fabricated. The void spaces of pores with different sizes and morphology generates Bragg diffraction and the wavelength of the peak changes when rebinding of target molecules happens. An optical signal via a change in Bragg diffraction is produced to detect atrazine. This optical property gives us an appropriate method of transducing recognition of target to a sensible signal.

Silica particle colloidal crystals were obtained by vertical self-assembly from 0.1% volume fraction suspensions in ethanol at 50°C. MIPs were fabricated with acrylic acid as the functional monomer, ethylene glycol dimethacrylate as crosslinker and UV-polymerization. The films, approximately 5 mm thin, were supported on polymethylmethacrylate (PMMA) slides.

Acrylic acid (AA), ethyleneglycol dimethacrylate (EGDMA), and azobisisobutyronitrile (AIBN) have been used as monomer, crosslinker, and initiator in polymerization precursor, respectively. Target molecules were also added to polymerization precure at 1:80 molar ratio to monomers. Then, a PMMA plastic slides on each sides of glass slides with depositions were firmly held together while immersed in polymerization precursor from one end. The polymerization precursor infiltrated through sandwich structure. Then, sandwich structure was held before UV light (365 nm) for 3 hours for the polymerization to take place. Inverse opal was created within polymer film by soaking PMMA slides in hydrofluoric acid (5%) and the silica particles were etched away, leaving empty cavities in the polymer. Finally, the target molecules were removed from film by washing with acetic acid and ethanol with ratio of 1:9. The MIPs were washed 3 times, for 15 minutes each time, using a shaker to enhance efficiency. Since detecting the metabolites is as essential as atrazine itself, MIPs were fabricated to detect DEA and DIA following the same steps.

Colloidal crystals were imaged by a FEI Quanta 600 FEG scanning electron microscopy to characterize the morphology of the crystal structure and the number of layers.

UV-Vis spectrophotometry was used to measure film surface reflectance to verify any swelling or shrinkage within the hydrogel. The reflectance spectra were recorded before and after exposure to clean water or atrazine over a wavelength range of 200–800 nm, using a benchtop double-beam UV–visible spectrophotometer equipped with a specular reflection accessory at a fixed angle of 30°.

Kinetic tests were conducted to investigate the time required for MIPs to reach equilibrium with a solution. Reflectance spectra were measured in specific time windows until the peak wavelength did not show any further changes. In kinetics test, the MIPs took 20 minutes to reach equilibrium, meaning it is necessary to incubate MIPs in unknown solutions for 20 minutes for the precise quantification of atrazine concentration.

The reflectance spectra of MIPs incubated in different solutions are illustrated in Figure 1. Changes in peak wavelength are a sign of swelling or shrinkage in hydrogel; in other words, adsorption of target molecules in binding sites of the polymer. The immersion of MIPs in solutions with higher concentration of atrazine changed the peak wavelength of the reflectance spectrum, as shown in Figure 1. In each step, peak wavelength increases in comparison with last more diluted solution which is a result of rebinding of target molecules in void binding sites in the structure of hydrogel. This target rebinding forms changes in porous structure of film, yielding different responses in reflectance spectrum each step. Then, based on this behavior and saturation time (20 minutes), MIPs were incubated in standard solutions with specific concentrations for 20 minutes with a gentle shaking and their reflectance were measure each step. The standard solutions, validated by LCMS, were used for measuring peak wavelength shifts for MIPs and to generate a calibration curve for quantifying unknown solution concentration of atrazine based on changes in wavelength shift.

The protocol for calibrating metabolites MIPs has been the same. Standard solutions validated by LCMS have been prepared and used to generate calibration curves by incubating metabolites MIPs in those standard solutions for 20 minutes on a shaker.

The fabricated sensor constitutes a simple, fast, and efficient tools to detect atrazine and metabolites in wetlands and also, quantify their concentration in order to make sure it does not go far higher than MCL. The thin film MIP-based sensors have a dynamic range of 0.14-12.3 ppb and calibration curves have been generated incubating MIPs in standard solutions with different concentrations.

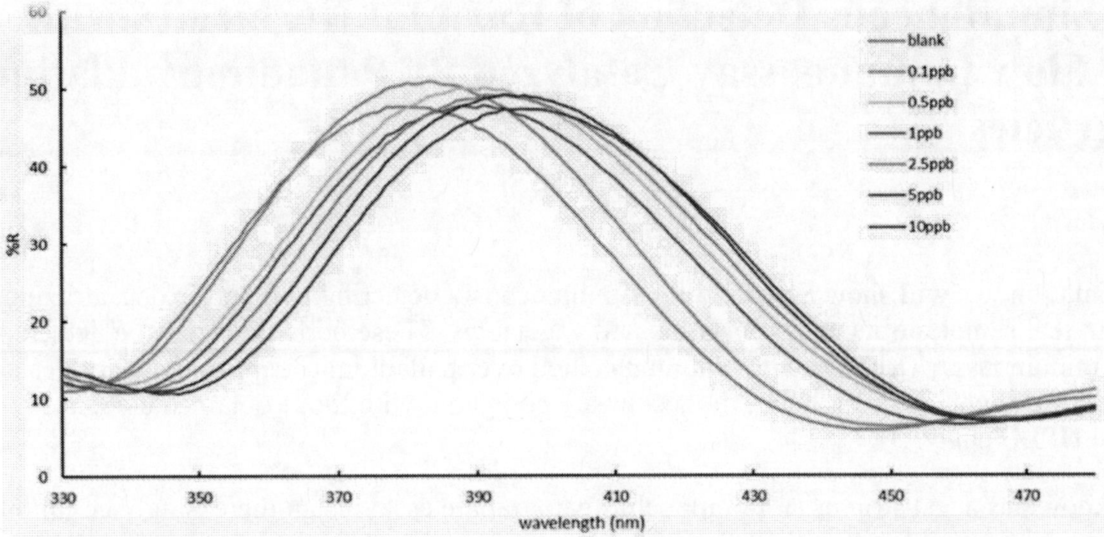

Figure 1. Optical responses of MIP to atrazine in a series of concentration.

598049 Simultaneous Detection of Harmful Herbicides with Lateral Flow Immunoassay Catalyzed By Palladium@Platinum Nanoparticles

Abstract:

In this presentation, we will show a method for simultaneously detecting two herbicides, atrazine and acetochlor, in real samples obtained from fruits and vegetables. These herbicides may be detected by lateral flow immunoassay (LFIA) using platinum coated over palladium (Pd@Pt) nanoparticles as a signal-amplifying agent. For validation, the results are compared with those obtained using a conventional HPLC analysis.

LFIA, also known as a test strip or an immunochromatographic assay, is an increasingly common immunoanalytical method for on-site screening because it is rapid, portable, inexpensive, and reagents are stable. LFIA has been applied in diverse areas such as medical diagnostics, drugs-of-abuse testing, environmental monitoring, and food safety. This study involves the use of LFIA for environmental monitoring and food safety, focusing especially on the detection of herbicides, since the capability of detecting these chemicals in the environment and in foods is crucial due to their potential hazards, from skin irritation to cancer. Atrazine is one of the most widely used herbicides in the United States, with 70 million pounds applied annually, as reported by the United State Geological Survey (USGS); however, the Environmental Protection Agency (EPA) has set the maximum contamination level in drinking water at 3 ng/mL for atrazine, and several studies have reported that atrazine has the potential to impact reproductive hormone levels in humans, rats, and amphibia. Acetochlor is another widely used herbicide, with an estimated 35 million pounds applied per year, as reported by the USGS; it is sometimes used instead of atrazine because of its superior biodegradability and higher maximum contamination level for drinking water, which is set at 20 ng/mL according to the Minnesota Department of Health. However, acetochlor is sited for its potential toxicity in rats and humans and has been noted by the EPA as having "Suggestive Evidence of Carcinogenic Potential."

Plans are underway to apply our previously developed dual-LFIA method, which employs a triangular sample pad combined with two test strips, to detect atrazine and acetochlor simultaneously in real samples of fruits and vegetables. We expect to be able to detect two herbicides near or below the maximum contamination levels set by the EPA and obtain recoveries between 80 – 120 %. The mesoporous Pd@Pt nanoparticle was used as the signal-amplifying agent because of its electrochemical activity as a peroxide conversion catalyst which allows for quantitative detection. Moreover, its high surface area increases its sensitivity, and its pH- and thermo-stability make it suitable for use with real samples. For validation, we detected the two herbicides simultaneously with HPLC and obtained a limit of detection of 10 ng/mL for atrazine and 50 ng/mL for acetochlor. We spiked apple and watermelon juices with 50, 80, and 100 ng/mL of atrazine and 100, 150, and 200 ng/mL of acetochlor, and obtained recoveries between 99.7 – 100 % when compared to an HPLC analysis. Laboratory experiments are planned for detecting the two herbicides in blended fruits, apples and grapes, and blended vegetables, corn and cabbage. We expect LODs near EPA limits and will report on the most recent progress. We expect our findings will suggest that innovations in efficiency, sensitivity, and stability are possible in tests for harmful chemicals in the environment and in food by use of the Pd@Pt nanoparticle enhanced immunoassays.

607001 Palladium Nanowires@Rgo@ZIF-8 Nanocomposite for High-Performance Hydrogen Sensors

Abstract:

Hydrogen forms a highly explosive mixture above 4% concentration in air. Therefore, fast detection of hydrogen in air is of high importance. Palladium has attracted attention as a material for room-temperature hydrogen sensing due to its unique ability to dissociate H_2 and form a hydride phase with higher electrical resistance than Pd metal under ambient conditions. However, the oxygen present in air often interferes with H_2 detection by generating moisture at the surface of Pd. Moreover, low-cost fabrication of Pd-nanowire sensors often produces sensors with non-monotonic response, due to competing effects of improved contact between nanowires and increased resistance of individual nanowires upon hydrogen uptake. Here, we report the fabrication of fast H_2 sensors using a novel ultra-thin and high-aspect-ratio Pd nanowires@reduced graphene oxide (PdNW@rGO) nanocomposite material coated with a layer of zeolite imidazole framework (ZIF-8). The rGO prevents direct contact between PdNWs and provides additional H_2-concentration dependent conduction channels, while the ZIF-8 layer limits oxygen transport to the NW surface. High-aspect-ratio Pd nanowires were synthesized in high yield by a hydrothermal method, followed by sonication with GO and subsequent reduction to form the PdNW@rGO nanocomposite. Pd nanowires of high-aspect-ratio increase the sensitivity of the sensor towards hydrogen while the layer of wrapped graphene prevents changes in NW contact resistance (i.e. creation of new conduction channels) due to volume expansion of PdNWs on hydrogen exposure. The ZIF-8 layer acts as a filter screening larger molecules like oxygen and nitrogen while allowing smaller molecules like H_2 to go through it. The hydrogen sensors on an interdigitated electrode (IDE) were exposed to concentrations (0.004-2%) of hydrogen diluted in synthetic air. The PdNWs@rGO@ZIF-8 hydrogen sensor showed a high-response of up to 4% resistance increase in air with a response time of 5 seconds to 1% H_2 in air and a lower limit of detection (LOD_{H2}) of 40 ppm. Relative to other fast Pd-based hydrogen sensors, the proposed solution-phase synthesis of PdNWs@rGO@ZIF-8 nanocomposite offers a combination of low cost and high-performance that is well-suited for applications such as fuel cell vehicles and hydrogen fueling stations.

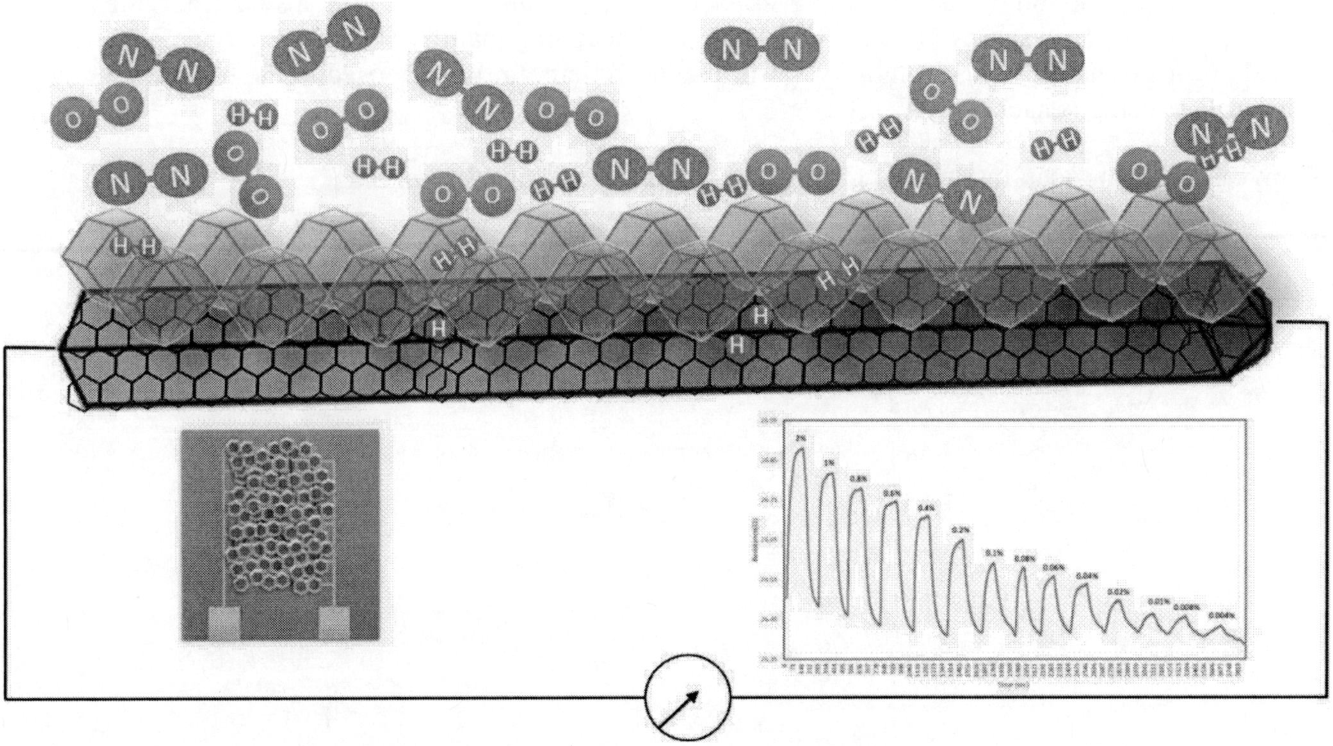

609139 Acoustic Wave Sensor Based on Plasticized Polymer Films for Detection of BTEX Compounds in Air

Abstract:

Although a variety of sensor technologies are available for environmental monitoring like photoionization, gas chromatography, mass spectroscopy, and e-noses sensors, chemical sensors based on piezoelectric transduction like the quartz crystal microbalance (QCM) and surface acoustic wave (SAW) device show promising results in terms of providing cost-effective real- time data, robustness, and low power requirements. Sensing films based on polymer-plasticizer films have been developed to detect the volatile organic compounds in the air, primarily benzene, ethylbenzene, toluene, and xylene (BTEX) using acoustic wave devices. This work shows for the first time how the selectivity and sensitivity of a glassy polymer film can be altered by adding a precise amount and type of plasticizer for BTEX detection.

To quantify the film saturation dynamics and understand the absorption of BTEX analyte molecules into the bulk of the sensing film, a diffusion study was incorporated to interpret the frequency vs time curve from a QCM. The sensor responses for individual analytes like benzene, toluene, etc. were collected and fit to exponentials which gives a characteristic value for response time, tau for individual analytes like benzene, toluene, ethylbenzene, and xylene which were found to be independent of the concentration of the analyte, thus helping in identifying the analyte. The model was also able to quantify individual analyte concentrations from the mixture based on the difference in tau values using a single polymer-plasticizer film on a single sensor. Several polymer/plasticizers materials and concentrations were screened to optimize the film for best BTEX sensing performance.

Quantifying BTEX concentrations from a mixture has always been challenging due to similarity in the chemical, physical, and molecular properties of the constituents. In this work, a single polymer-plasticizer film-based sensor has been employed to detect and quantify BTEX components in air based on sorption properties (solubility, partition coefficient, tau) as opposed to the traditional approach of using sensor arrays.

601716 Assessment of Residual Life of a Gas Filter Using Carbon-Nanotube Gas Sensors

Abstract:

A gas-filtration system filled with porous adsorbent is one of the most useful tools for reducing the amount of toxic volatile organic compounds released to the environment. However, it is challenging to assess the residual life of a filtration system and predict the time to replace the filter. In this study, we have shown how to evaluate the residual life of a gas-filtration system by embedding gas sensors into the system. Accordingly, chemiresistive gas sensors were fabricated via the deposition of patterned electrodes on a thin film of carbon nanotubes (CNTs). Multiple sensors were sequentially installed in the gas filter along the gas-flow direction. Dimethyl methylphosphonate, and ammonia were used to form a toxic gas model, and we observed the real-time changes in the resistance of the CNTs, which are highly sensitive to the adsorption of gas molecules because of their one-dimensional structure with high surface-to-volume ratio. The sensors responded sequentially as the toxic gases broke through the bed, and these responses could be correlated to the breakthrough curve of the toxic gases, obtained using gas chromatography and FT-IR. Here the properties of the CNT sensor can be potentially tuned by surface chemistry for the sensitive and selective detection of a target organic compound. Therefore, the approach reported herein can be useful for real-time assessment of the residual life of various gas-filtration systems.

601487 Ink Synthesis and Screen-Printing of Graphene Oxide Ink for Paper Based Electrochemical Sensors.

Abstract:

There has been a growing interest in making high performance electrode materials from graphene oxide (GO) to be used in electrochemical sensors, which has since gained popularity due to the excellent analytical performance, low-cost and simple properties linked to the devices. Herein, a method for fabricating highly conductive GO and GO@g-C3N4based ink to be used for screen printing electrochemical sensors. The physical properties of graphene oxide and GO@g-C3N4 were characterized by different measurements. Methods: Brunauer–Emmett–Teller (BET) surface area measurements, scanning electron microscopy (SEM), and Raman analysis, and GO was prepared using the modified Hummers' method. Furthermore, the viscosity, conductivity and surface tension of the printable GO and GO@g-C3N4 based ink were studied. Disposable screen-printed electrode electrochemical sensors were developed from GO-based inks for detecting heavy metals in effluent wastewater. Therefore it is expected that the GO-based inks will increase the sensitivity and selectivity of the electrode sensors.

605026 Effect of Surface Heterogeneity, Heat of Adsorption and Surface Area on the Characteristics of PANI-SnO_2 Based H_2s Gas Sensor

Abstract:

The emeraldine salt form of polyaniline (PANI) is a p-type organic semiconductor with polarons as charge carrier species[4] (figure 1 (a)). The protonation level of PANI changes upon exposure to a reducing or an oxidizing gas, which results as change in conductivity of the material[2]. Composites of the PANI with several metal oxides have been used to improve the sensor characteristics like sensitivity, selectivity, response time and limit of detection. SnO_2, one such metal oxide, makes a co-ordination with the N in PANI which results in a p-n junction between p-type polyaniline and n-type SnO_2[3]. Also, co-ordination with SnO_2 affects the protonated sites of polyaniline which introduces surface heterogeneity. The sensor characteristics are affected by surface heterogeneity as well as electrical properties of the material. Apart from this the surface area and morphology of the material also plays a major role in assisting the adsorption of the gas molecules on the sensor surface. In this work the effect of mixing the different amounts of SnO_2 in polyaniline on the H_2S sensing characteristics has been studied. The fractional change in electrical resistance of sensing element upon exposure to H_2S gas was recorded as the sensor response, and the surface site kinetics were calculated according to nature of the sensor response. The surface site kinetics were studied with Langmuir and Freundlich isotherms and the kinetics parameters were compared with the material properties such as heat of adsorption, surface area, and band gap.

PANI-SnO_2 composites were synthesized by solution casting method[3]. Four different wt. % (0, 5, 10 and 15) of SnO_2 nanoparticles (Sigma Aldrich, particle size < 100 nm) were added to 20 kDa molecular weight PANI emeraldine base (Sigma Aldrich) in 1 molar solution of camphorsulfonic acid (CSA) (Sigma Aldrich). The mixture was kept on stirring for 24 hours, during mixing the PANI gets protonated by CSA and SnO_2 forms co-ordination with N of PANI as shown in Figure 1 (b). After this reaction mixture was filtered, washed with DI and acetone, and dried. The composite powders were labeled as P-00, P-05, P-10, and P-15 according to increasing order of the SnO_2 loading.

The surface area of PANI decreased by addition of SnO_2, but the pore size was unaffected (table 1). The UV-Vis spectra showed π-π^* and polaron-π^* transitions at 350 nm and 450 nm respectively, and a free carrier tail which confirmed polaron form of polyaniline with expanded coil structure[5]. Band gap of the material was increased with 5 wt. % SnO_2 due to non-uniform dispersion of the n-type metal oxide. However, the band gap decreased on further addition of SnO_2 due to more co-ordination of Sn with PANI. The N1s XPS spectra confirmed amine, imine and N^+ type surface nitrogen moieties[6]. The percentage of active sites for each composite were approximated from N1s spectra.

The sensing elements were fabricated by drop casting the 10 wt. % slurry of composite powders in DI water over thermally deposited silver interdigitated electrodes on PET substrate. The sensor was placed in a 125 mL cubical chamber and nitrogen gas was flushed to make the atmosphere inert. Calculated amounts of H_2S gas were injected to make 10, 50 and 100 ppm concentrations inside the sensing chamber, and the electrical resistance of the sensing element was recorded by a digital multimeter (Keysight 34465a). The response time was calculated as the time taken by resistance in reaching the 63.2% of its saturation value. And the sensitivity was calculated in ppm^{-1} from the slope of response v/s H_2S concentration plot.

H_2S is a reducing gas and it also shows mild acidic behavior in presence of moisture[1], so it can protonate as well as deprotonate the PANI which result as change in the resistance. Since there was no moisture in sensing chamber, so the reducing behavior of H_2S caused deprotonation of the PANI, which increased resistance of the sensing element upon exposure to H_2S. For these sensors (P-00, P-05, and P-15) N^+ sites were considered as active site for chemisorption and surface site kinetics was evaluated from equation (1). However, the sensor P-10 had more uniform dispersion of SnO_2 in PANI, so there was more co-ordination of SnO_2 with PANI. In this case acidic behavior of H_2S caused protonation of PANI, and imine N sites were considered as active site of chemisorption. The chemisorption kinetics of surface sites was calculated from equation (2) for this sensor.

The response time of PANI sensor was improved on adding SnO_2 but the sensitivity was decreased (Figure 2(a)). Sensor P-10 show fastest response, but the sensitivity was decreased to 0.003 ppm^{-1}. Also, the reaction chemistry for P-10 sensor was different because the decreasing resistance response indicated protonation of PANI showing the acidic behavior of H_2S. The shorter response time for acidic behavior of the H_2S has been reported previously[1]. For other sensors (P-00, P-05, and P-15), the response time decreased with increase in band gap of the material, such behavior shows the importance of electrical properties of the material. The surface site kinetics (figure 2 (b)) show exponentially increasing behavior fitted with Langmuir kinetics (equation 3) and the adsorption isotherm show Freundlich (equation 4) behavior indicating heterogeneity in the surface sites of the sensing material. The adsorption coefficient α decreased with heat of adsorption for PANI-SnO_2 composites (figure 2 (c)). However, pristine PANI does not follow this trend due to more surface area (Table 1).

A high surface area with more active surface sites, and high heat of adsorption is desired for good sensitivity. A high band gap will give better response time of the sensor. However, the nature of surface chemistry will also play an important role in the sensor response characteristics. To design a PANI-SnO_2 composite based H_2S sensor, the environmental conditions (humidity level) of the sensor application area and synthesis procedure for more co-ordination of SnO_2 with PANI should be focused to promote role of the surface chemical reaction.

References

1. **Agbor, N.E., M.C. Petty, and A.P. Monkman**, "Polyaniline thin films for gas sensing," *Sensors and Actuators B: Chemical*, **28** (3), pp. 173–179 (1995).
2. **Hu, H., M. Trejo, M.. Nicho, J.. Saniger, and A. Garcıa-Valenzuela**, "Adsorption kinetics of optochemical NH3 gas sensing with semiconductor polyaniline films," *Sensors and Actuators B: Chemical*, **82** (1), pp. 14–23 (2002).
3. **Khuspe, G.D., S.T. Navale, M.A. Chougule, *et al.***, "Facile method of synthesis of polyaniline-SnO2 hybrid nanocomposites: Microstructural, optical and electrical transport properties," *Synthetic Metals*, **178**, pp. 1–9 (2013).
4. **Stafström, S., J.L. Brédas, A.J. Epstein, *et al.***, "Polaron lattice in highly conducting polyaniline: Theoretical and optical studies," *Physical Review Letters*, **59** (13), pp. 1464–1467 (1987).
5. **Xia, Y., J.M. Wiesinger, A.G. MacDiarmid, and A.J. Epstein**, "Camphorsulfonic Acid Fully Doped Polyaniline Emeraldine Salt: Conformations in Different Solvents Studied by an Ultraviolet/Visible/Near-Infrared Spectroscopic Method," *Chemistry of Materials*, **7** (3), pp. 443–445 (1995).
6. **Xiong, P., H. Huang, and X. Wang**, "Design and synthesis of ternary cobalt ferrite/graphene/polyaniline hierarchical nanocomposites for high-performance supercapacitors," *Journal of Power Sources*, **245**, pp. 937–946 (2014).

a) b)

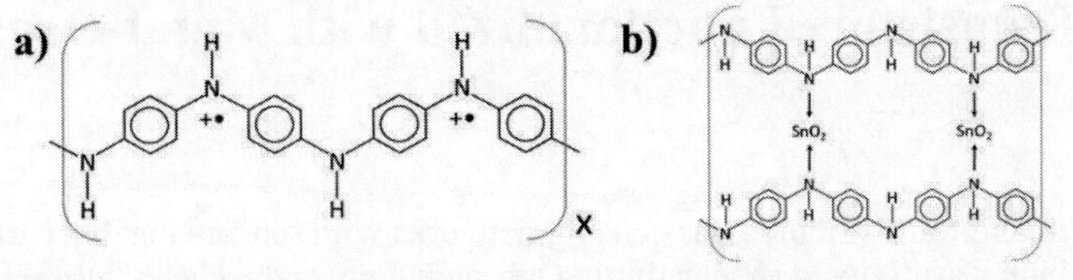

Figure 1. a) Polaron structure of Polyaniline Emeraldine Salt, b) SnO_2 co-ordination with PANI

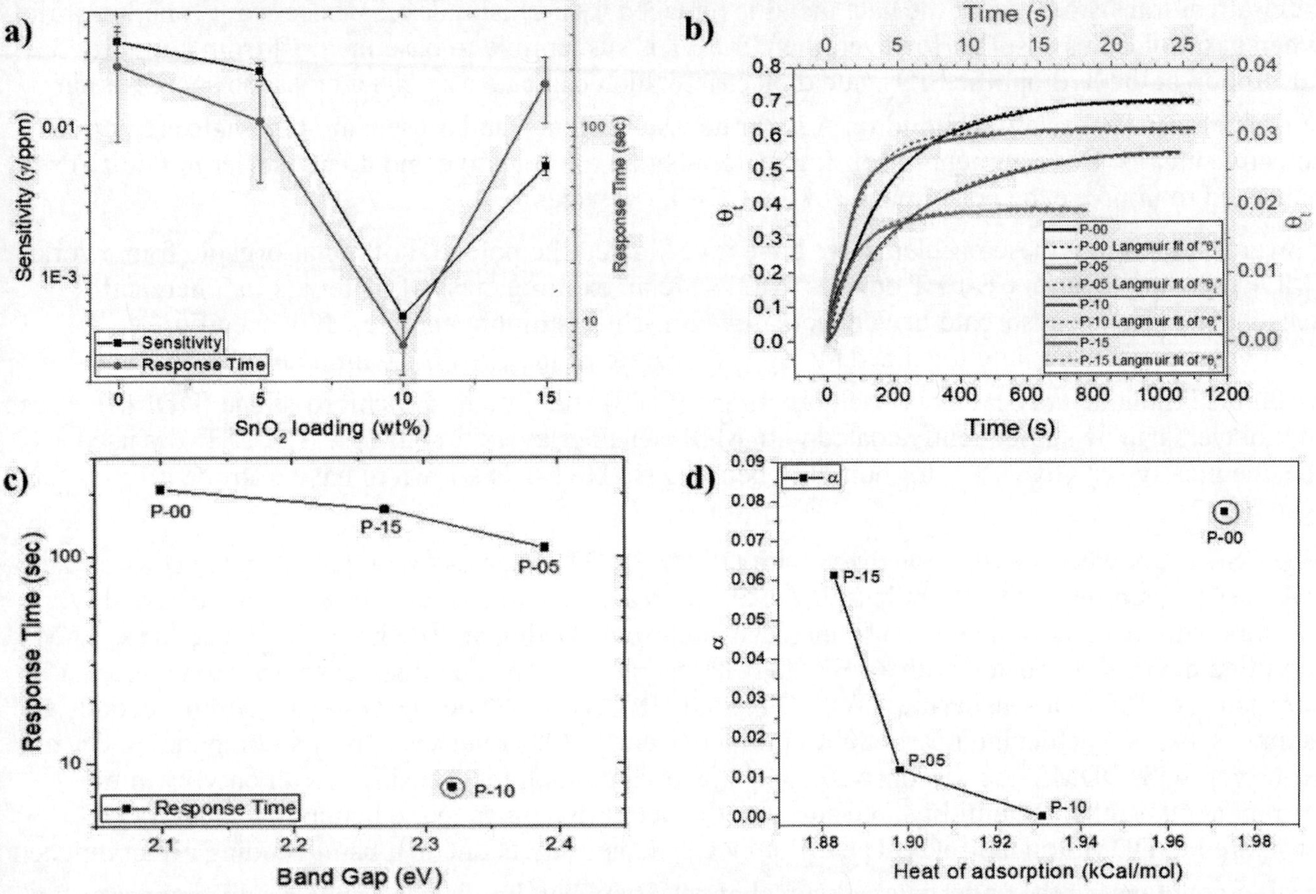

Figure 2. a) Sensitivity and response time of PANI-SnO_2 based H_2S sensors. b) Surface site kinetics for 50 ppm H_2S environment. c) Freundlich isotherm coefficient (α) and heat of adsorption (Q) correlation

Equations:

$$\theta(t) = \frac{\gamma(t)}{1+\gamma(t)} \quad (1)$$

$$\theta(t) = \frac{\gamma(t)}{1-\gamma(t)} \quad (2)$$

$$\theta(t) = \frac{N^*}{N_t} = \theta_0\{1-\exp(-\nu t)\} = \frac{P}{P+b}\{1-\exp(-\nu t)\} \quad (3)$$

$$\theta_0 = \propto P^{1/n} \quad (4)$$

Table 1: BET Surface area, heat of adsorption, and BJH pore size of the PANI-SnO_2 Composites

Material	BET surface Area (m^2/g)	Heat of adsorption (kCal/mol)	BJH Pore size (Å)
P-00	63.53	1.97	14.2
P-05	27.95	1.89	14.2
P-10	28.71	1.93	14.2
P-15	29.52	1.88	16.7

607925 Effects of Alkane Dielectrics in Chemically-Sensitive Field-Effect Transistors Functionalized with Metal-Organic Frameworks

Abstract:

There is a growing demand for distributed gas-sensing networks with sensors that have ultra-low-power consumption, high sensitivity, and selectivity. One such device, the chemically sensitive field-effect transistor (CS-FET), meets these criteria(1). The CS-FET is structurally similar to a standard field-effect transistor, except the gate metal is replaced by a sensing layer whose work function shifts when gas molecules adsorb. However, the CS-FET is susceptible to baseline drift from humidity adsorption at the hydrophilic SiO_2 gate dielectric, which can cause a 50% increase in source-drain current at just 40% relative humidity. A separate issue is that standard sensing materials are non-selective metals. Development of sensing materials that are selective and do not suffer humidity drift is needed to improve the performance of the CS-FET devices.

Towards addressing these problems, we have investigated the potential of metal-organic framework (MOF) sensing layer in CS-FET devices. MOFs are an exciting class of materials for chemical sensing because the adsorbate-driven work-function-shift is highly analyte-MOF specific. Furthermore, we have functionalized the SiO_2 dielectric of the CS-FET with hydrophobic self-assembled monolayers octadecyl-trichloro-silane (OTS) and dimethyl-dichloro-silane (DDMS). These monolayers can be subsequently coated with MOF sensing layers; in our case, HKUST-1 is used as the chemically sensitive layer for humidity because HKUST-1 is known to have a strong affinity for water(2).

The CS-FET devices are functionalized with DDMS or OTS. Sensors with just OTS have nearly no response to humidity, with about less than a 5% increase in source-drain current at 40% humidity. Sensors with just DDMS have a 10% increase in current starting at 40% humidity. Once these SAM-modified devices are coated with an HKUST-1 sensing film, we find distinct sensing dynamics and, surprisingly, different sensitivities. With OTS and HKUST-1, the device responds within seconds and is quite sensitive, achieving nearly 800% initial current at 40% relative humidity compared to dry air. However, with DDMS, the response is much slower and much less sensitive, with only 2x initial current at 40% relative humidity. We attribute the better dynamics to the better coverage of OTS compared to DDMS on the SiO_2. The difference in sensitivity is due to a band bending effect induced by the monolayer in the underlying silicon channel. Band bending is confirmed by differences in the electrical characteristics of the devices.

Inspired by these results we have begun to work on sensors that replace the SiO_2 dielectric completely with an alkane dielectric, 1-octadecene, that grows directly on silicon. Preliminary results show that a device with just this monolayer has even a weaker response to humidity than OTS. We plan to evaluate this monolayer as a dielectric for an HKUST-1-based humidity sensor and evaluate the effect of the monolayer on the electrical properties of the CS-FETs.

References:

1. H. M. Fahad, N. Gupta, R. Han, S. B. Desai, A. Javey, Highly Sensitive Bulk Silicon Chemical Sensors with Sub-5 Nanometer Thin Charge Inversion Layers. *ACS Nano*. **12**, 2948–2954 (2018).
2. D. W. Gardner, X. Gao, H. M. Fahad, A.-T. Yang, S. He, A. Javey, C. Carraro, R. Maboudian, Transistor-based work function measurement of metal-organic frameworks for ultra-low-power, rationally designed chemical sensors. *Chem. – A Eur. J.* (2019), doi:10.1002/chem.201902483.

603892 (Invited) Skin-like Wearable and Implantable Sensors

Abstract:

In this talk, I will present our recent work on BodyNet concept for stretchable wearable and implantable sensors. Examples include, full body sensing, blood flow sensor and morphing electronics - neurostimulator.

600113 An Adhesion-Based Anti-Corrosion Strategy for Wearable Electrochemical Sensing and System Integration

Abstract:

The large-scale deployment of biomonitoring sensors through integration in wearable consumer electronics enables population-level health and wellness monitoring, and thus can play a critical role in transforming personalized medicine. Currently, physical sensors have been widely incorporated within commercialized wearable platforms, but they cannot capture biomarker information that lie in molecular levels and provide highly specific information about the body's dynamic chemistry. Electrochemical biosensing interfaces can be positioned to target biomarker molecules in a wearable format, as they can be constructed on miniaturized/flexible footprints/substrates and can convert (bio)chemical signals into electrical signals measurable by electronic readouts. The contact pads on the printed circuit board (PCB), which is a core component of wearable consumer electronics, can directly serve as electrodes and as a substrate for electrochemical sensors. However, we found that the commercialized PCB metal contact pads (*e.g.*, gold) were prone to corrosion and unsuitable for electrochemical modifications in solution environments. Additionally, electrochemical biosensors often are not reusable and need to be replaced frequently, while PCB units are not positioned for disposable applications.

To this end, we devise an adhesion-based anti-corrosion strategy for PCB-interfaced electrochemical sensing, which enables seamless sensor-system integration. The strategy is based on decoupling the sensing interface and PCB metal contact pads by an intermediary ultrathin (~50 μm), vertical conductive, and double-sided adhesive film. This strategy has dual benefits of making electrochemical sensors disposable and insulating the contact pads from sounding solutions. We first establish the suitability of the adhesion film to serve as an electrode and as an interconnection film simultaneously. One side of the double-sided adhesive film was patterned with noble metal electrodes, followed by the modifications with recognition layers, such as permselective membrane and enzymes, to render biosensing functionality. The electrochemical biosensors were transferred and adhered to the PCB contact pads with the aid of water-soluble tape. As compared with bare PCB contact pads, this fabrication strategy allows for realizing a corrosion-resistant and stable electrochemical biosensing interface on PCB for operation in biofluid (*e.g.*, sweat). Amperometry and open circuit potential characterization experiments validated the electrochemical corrosion resistance properties in saline solution environments. The excellent anti-corrosion attributes of the modified film could be ascribed to: (1) the polymer glues on both sides of the adhesive film acting as corrosion-protective layers; (2) bypassing the need for oxidation prone-adhesion metal (*e.g.*, chromium), achieved by e-beam evaporation (at the noble metal patterning step); and (3) the electrodeposited permselective membrane (here, phenylenediamine), which has been reported to presented anti-corrosion properties. We specifically tailored the enzymatic layer of the electrochemical interface to target biomarker molecules such as glucose, lactate, and choline as example of informative metabolites and nutrients. By modifying different (bio)chemical layers various other molecules can be targeted. In this way, our strategy serves as a foundation for electrochemical sensor development and integration with wearable consumer electronics.

607021 Rapid Screening Platform for the Isotopic Determination of Radionuclides in Water

Abstract:

Traditional approaches for the detection of trace-level radioisotopes in water require lengthy offsite sample preparations and do not lend themselves to rapid screening. Therefore, a novel sensor platform that combines onsite purification, concentration, and isotopic screening with a fieldable detection system will be an invaluable tool.

In this presentation, we will discuss findings on the development of a sensor platform based on reactive polymer film coatings for the rapid isotopic screening of waterborne radionuclides. Nanothin coatings were prepared by grafting uranium-selective polymers from polyethersulfone (PES) surfaces via UV-initiated polymerization, and by introducing uranium-selective functional groups to polyacrylonitrile (PAN) surfaces by chemical reaction. Ellipsometry was used to study film growth kinetics on PES films. X-ray photoelectron spectroscopy of modified PAN films revealed conversion of nitrile groups to amidoxime groups to be as high as 40% and showed that the extent and depth of reaction could be varied precisely. Static uptake experiments with solutions of depleted uranium spiked with ^{233}U were conducted to determine uranium binding capacities and kinetics of the modified polymer films at different pH values from 4 to 8. Sorption isotherm data were fitted to the Langmuir model, and the highest sorption capacities were obtained at pH 6 for modified PAN (M-PAN) and PES (M-PES) films. Capacities at pH 4 and 8 were lower and could be explained by differences in sorption mechanisms. Uranium batch uptake kinetics followed a pseudo-second order rate model. Alpha spectroscopy pulse height spectra were analyzed to study the role of selective layer film thickness on peak energy resolution. Excellent resolution with full width at half maximum values from 29 to 41 keV were recorded for M-PAN film and from 26 to 45 keV for M-PES film. Whereas uranium uptake increased with selective layer film thickness and varied with polymer chemistry/extent of modification, the peak energy resolution was independent of layer thickness and polymer chemistry within the experimental measurement uncertainties.

Results from this work were used to modify PES and PAN ultrafiltration membranes with the same experimental conditions to develop uranium selective membranes. High-throughput uptake experiments were performed to study the effect of residence time on detection efficiency and isotopic resolution. This membrane-based detection method enables the rapid, fieldable screening analysis of radionuclides in water for environmental monitoring studies and nuclear forensics investigations.

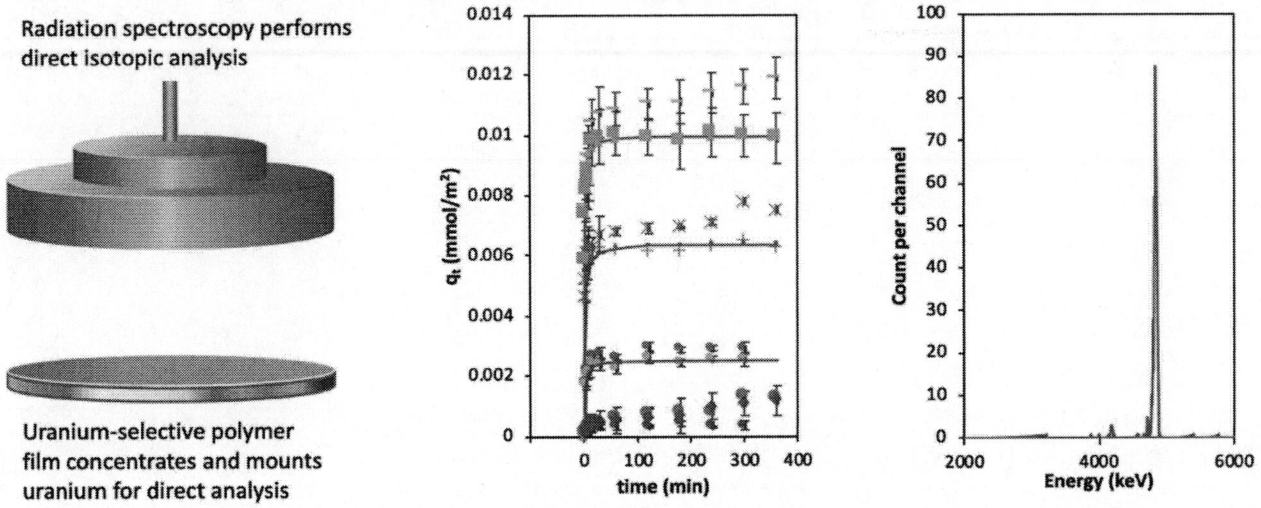

600875 Electrochemical Detection of Environmental Pollutants

Abstract:

Many small molecule pollutants interact with human hormonal systems. These compounds, known as endocrine disruptors, are ubiquitous, as they are found in plastics, medications, and pesticides. Such chemicals are especially problematic because they have been linked to severe health problems, including cancer, infertility, and diabetes. We have developed two complementary electrochemical methods to rapidly detect and quantify these compounds. One technique involves the specific monitoring of Bisphenol A (BPA) using a DNA aptamer immobilized on an electrode. The second method involves the measurement of the biological activity of chemically dissimilar endocrine disruptors using a human hormone receptor as a component of an electrochemical sandwich assay. These electrochemical platforms require no specialized skills to implement and have enabled very sensitive detection (sub-ppb levels) of these environmental pollutants from complex solutions, such as baby formula.

604144 Fabrication of High-Resolution Graphene-Based Flexible Electronics Via Polymer Casting and Microfluidic Approaches

Abstract:

Graphene-based flexible electronic devices for sensor applications has been receiving growing interest due to unique properties of graphene.[1] Various fabrication methods, such as photolithography, printing and chemical vapor deposition (CVD), has been used to produce high resolution and low feature size graphene-based devices.[2, 3] However, most of these methods are complex and require multiple processing steps (i.e. stamping, vacuum drying and etching) as well as additional post treatments (i.e. high temperature (300-1000 °C) baking or laser annealing) that can thermally or chemically degrade polymer-based substrates limiting substrate material selection.[2, 3] In addition, these methods can only be applied to 2D planar substrates and are not feasible to fabricate 3D circuits and some of them, such as sticky/adhesive tape peeling method, require high amounts of graphene consumption.[4] Despite the recent progress in the field, there is still a need to develop novel fabrication approaches that can address the current limitations of the existing methods.

In this study, we present two flexible electronics device fabrication methods based on simple polymer casting and microfluidics approaches.[5, 6] The first method focuses on simple polymer casting based graphene transfer at room temperature that does not require any post-processing. Briefly, our method consists of two main steps; (i) the formation of graphene patterns/films on substrates/molds via conventional methods such as CVD, channel filling or ink-jet printing and (ii) direct casting of target substrate polymer solution on the substrates/molds with graphene micropatterns and direct graphene transfer to the target substrate via peeling off upon drying and film formation. This method simply relies on the differences in the surface energies and adhesive forces between the graphene/mold and graphene/target polymer substrate.[5] Alternatively, the second method concentrates on a novel microfluidic approach to fabricate high resolution and low feature size graphene-based circuits/devices.[6] This method involves controlled and selective filling of microchannels on substrate surfaces with a conductive binder-free graphene nanoplatelets (GNP) solution. The ethanol-thermal reaction of GNP solution at low temperatures (~75 °C) prior to microchannel filling (as called "pre-heating") potentially reduces and homogenize GNP, which in turn enhances conductivity and reduces sheet resistance (down to ~0.05 kΩ sq^{-1}). Therefore, harsh post-processing of GNP is eliminated, and room temperature fabrication is enabled, which allows the use of degradable substrates, including biodegradable and natural polymers. This simple process is followed by selective administration of conductive GNP solution to the predetermined microfluidic channels on flexible substrates in a controlled manner via a syringe pump to create high resolution and low feature size graphene circuits/devices. This microfluidic approach can also be used to create 3D circuits with different geometries, using combined approaches with 3D printing, which is difficult to obtain with other methods (i.e. inkjet, screen or aerosol printing) that work mainly with 2D planar substrates.[6]

We demonstrated that using these fabrication methods, we obtained high resolution and low feature size graphene circuits with different geometries, dimensions and 3D superficial microstructures. We investigated the effect of temperature and microfluidic parameters (such as concentration, volume and flow rate) as well as the type of substrate materials on the graphene structure and formed micropatterns. The XPS, Raman, SEM and TEM results validated the structural changes of graphene along with confirming the stable presence of graphene patterns on the target substrates.[5, 6] The feature sizes of the graphene patterns could range from a few micrometers (down to ~15 μm in width and ~5 μm in depth) to a few millimeters with very small amounts of GNP used (~2.5 mg of graphene to obtain ~0.1 kΩ sq^{-1} of sheet resistance for 1 cm^2). [5, 6] The generated graphene patterns demonstrated significant stability after multiple washing and bending cycles. These methods were also implemented using other conductive inks, such as conductive graphene-silver based composite ink.[5, 6] Room temperature processing also provided precise control of 3D microstructural and mechanical properties (such as film porosity, pore size, elasticity etc.) of the target substrate materials, as demonstrated by SEM images.[5, 6] Using these methods, we successfully fabricated pressure sensors and near field communication antennas. [5, 6] Our results demonstrated that the devices worked with high sensitivity and stability.

In conclusion, the circuits/devices fabricated using this method can easily be implemented as wearable electronic, sensors and batteries, robotic components or motion detectors. In a broader sense, these simple, environmentally friendly and low-cost methods could pave the way for the fabrication of 2D and 3D electronic circuits/devices on different degradable flexible substrates using various graphene-based conductive inks.

References:

1. Jang, H.; Park, Y. J.; Chen, X.; Das, T.; Kim, M. S.; Ahn, J. H., Graphene-Based Flexible and Stretchable Electronics. *Advanced Materials* **2016,** *28* (22), 4184-4202.
2. Das, S. R.; Nian, Q.; Cargill, A. A.; Hondred, J. A.; Ding, S.; Saei, M.; Cheng, G. J.; Claussen, J. C., 3D nanostructured inkjet printed graphene via UV-pulsed laser irradiation enables paper-based electronics and electrochemical devices. *Nanoscale* **2016,** *8* (35), 15870-15879.
3. Das, S. R.; Srinivasan, S.; Stromberg, L. R.; He, Q.; Garland, N.; Straszheim, W. E.; Ajayan, P. M.; Balasubramanian, G.; Claussen, J. C., Superhydrophobic inkjet printed flexible graphene circuits via direct-pulsed laser writing. *Nanoscale* **2017,** *9* (48), 19058-19065.
4. Oren, S.; Ceylan, H.; Schnable Patrick, S.; Dong, L., High-Resolution Patterning and Transferring of Graphene-Based Nanomaterials onto Tape toward Roll-to-Roll Production of Tape-Based Wearable Sensors. *Advanced Materials Technologies* **2017,** *2* (12), 1700223.
5. Uz, M.; Jackson, K.; Donta, M. S.; Jung, J.; Lentner, M. T.; Hondred, J. A.; Claussen, J. C.; Mallapragada, S. K., Fabrication of High-resolution Graphene-based Flexible Electronics via Polymer Casting. *Scientific Reports* **2019,** *9* (1), 10595.
6. Uz, M.; Lentner, M. T.; Jackson, K.; Donta, M. S.; Jung, J.; Hondred, J.; Mach, E.; Claussen, J.; Mallapragada, S. K., Fabrication of Two-Dimensional and Three-Dimensional High-Resolution Binder-Free Graphene Circuits Using a Microfluidic Approach for Sensor Applications. *ACS Applied Materials & Interfaces* **2020,** *12* (11), 13529-13539.

610704 Tetracosane Functionalized Nanostructure Titanium Oxide Sensors for Screening Pneumonia from Breath

Abstract:

Volatile organic compounds (VOCs) are present in the environment and human bodies. Ethanol, acetone, and isoprene are examples of VOCs in the healthy human body, specifically in breath[1]. Diseases such as pneumonia have VOCs associated with it in breath[2]. Pneumonia is the cause of 16% death of children under five globaly[3]. In the US, pneumonia is the first common reason of hospitalization in children and the second common reason for adults' admission after women giving birth[3]. Therefore, there is a need for an inexpensive, noninvasive, and quick pneumonia screening method that can simply be performed in clinics on the first patient visit. Patients with pneumonia have been shown to have breath profiles that include unique VOCs that derive from the biological pathogen and perturbations in the health of a patient. The presence of heptane and increase in the amount of ethanol are examples of VOCs associated with pneumonia[2]. Developing a device to measure the presence of these VOCs in breath could be used as a screening method to detect pneumonia. In this work, a suitable molecule is functionalized on a titanium oxide platform to give an electrochemical response in the presence of pneumonia VOCs.

The sensor synthesis process consists of anodization of titanium oxide nanotubes and annealing the anodized self-ordered nanotubes under oxygen. Titanium oxide nanotubes provide an adequate conductivity and have an amorphous surface area suitable for external coatings. As mentioned earlier, ethanol is at a higher concentration in the breath of patient's suffering from pneumonia compared healthy human's breath[2]. Titanium oxide nanostructure at given voltages responds to the presence of ethanol by showing current changes. But for the purpose of detecting pneumonia, a more specific sensor is needed. To design a sensor that also responds to heptane, titanium oxide nanotubes were functionalized with tetracosane. Tan et al.[4] showed that when alkane gets mixed with tetracosane on a conductive surface. At a constant voltage of 1.5V, the tetracosane functionalized titanium oxide sensor performs as a chemiresistive sensor. It responds to heptane and ethanol in the gas phase by showing current changes. Table (1) shows the current response for a tetracosane functionalized titanium oxide sensor to ethanol, heptane and humid air exposure. Due to the presence of humidity in the human breath, the sensor was exposed to controlled humid samples. When heptane, humid air, and ethanol were exposed to the functionalized sensor, the current increased to different magnitudes (figure 1). Room air was bubbled through a solution of the VOC of interest and exposed to the sensor that was placed on PCB board. The current change magnitude and the response time of the sensor varied for each exposure. The sensor's response to humid air and ethanol was instant, whereas the coated tetracosane sensor responded to heptane exposure after a minimum of 40 seconds. The sensor's current change to ethanol exposure was a minimum of three orders of magnitude higher than heptane in the first 100 seconds of sample exposure. The tetracosane functionalized sensor shows a different response to each heptane, ethanol, and humid air in gas phase. The sensor's response to different concentrations of heptane and ethanol separately and mixed together in mimicked breath samples will be presented.

References:

[1] J. D. Fenske and S. E. Paulson, "Human breath emissions of VOCs," *J. Air Waste Manag. Assoc.*, vol. 49, no. 5, pp. 594–598, 1999.

[2] R. Schnabel et al., "Analysis of volatile organic compounds in exhaled breath to diagnose ventilator-associated pneumonia," Sci. Rep., vol. 5, no. July, pp. 1–10, 2015.

[3] N.A., "Top 20 Pneumonia Facts 2018," Am. Thorac. Soc., p. 2018.

[4] Jiunn-Liang Tan; Zheng-Xin Yong; Chong-Kin Liam, Using a chemiresistor-based alkane sensor to distinguish exhaled breaths of lung cancer patients from subjects with no lung cancer. *Journal of Thoracic Disease* 2016, 10.2103

	Heptane	Water	Heptane	Ethanol
Current Change (uA)	2.19e-2	3.04e-2	4.53e-2	7.41e+1

Table (1), Sensor's current changes when heptane, water, and ethanol was exposed to it, at 1.5V.

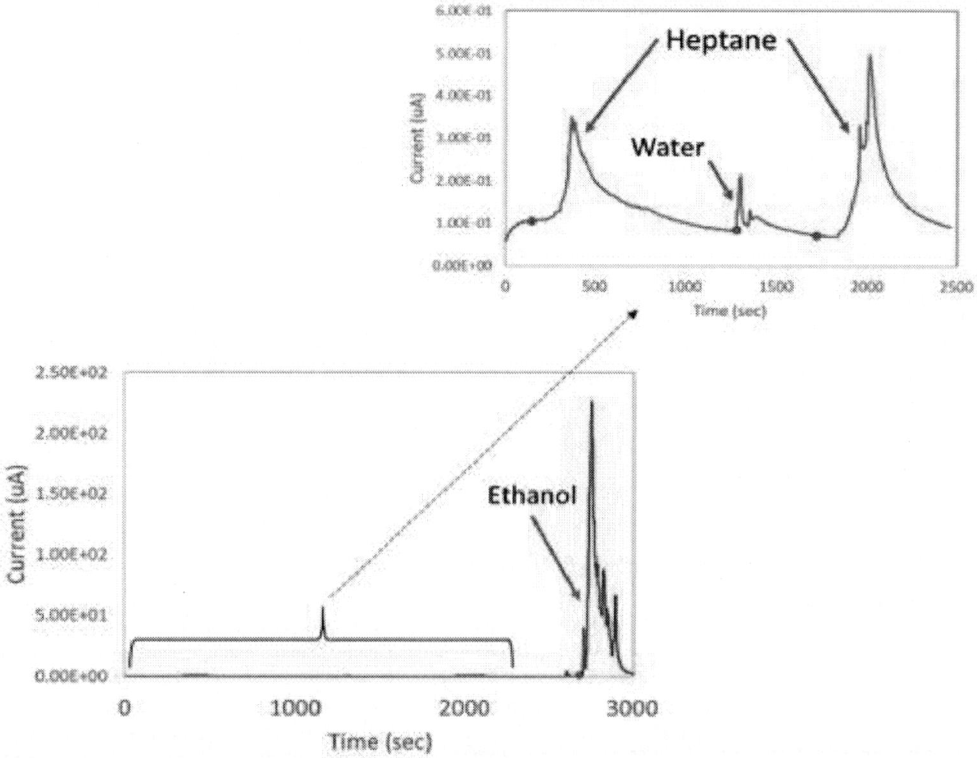

Figure (1): An example of tetracosane sensor response to Heptane, water and ethanol. The Blue dots are the starting time the sensor was exposed to the sample.

603401 Fabrication of Enzymatic Electrochemical Biosensors for Real-Time Nicotine Detection

Abstract:

Biosensors provide an output signal based upon a successful recognition of an analyte. Key to this process in oxidase-based biosensor is analyte recognition by an enzyme and subsequent signal transduction. As such, the sensor design process must successfully address a series of requirements associated with the interface between the enzyme and an electrode. Here we present the development, assessment, and optimization of a nicotine sensor based on a nicotine catabolizing redox enzyme and Prussian Blue coated screen-printed electrodes to afford a chronoamperometric response proportional to nicotine. An optimized polymeric matrix immobilization method maintains enzymatic activity on the mediator-coated screen-printed electrode and enables quantitative detection of nicotine. The resulting biosensor detects nicotine in urine with high sensitivity with a limit of detection of 34 μM over the range of 0-200 μM.

611271 Point-of-Care Fungal Pathogen Identification Using Surface Acoustic Waves Coupled with Metal-Enhanced Fluorescence

Abstract:

Corneal ulcers, typically caused by infection, are among the leading causes of vision loss and require prompt intervention. Without effective treatment, permanent vision impairment[i] and more serious infections[ii] can result. Rather than spending days culturing the pathogen to determine whether the infection is bacterial or fungal, antibiotics are typically applied as an empiric treatment[iii]; as the incidence of fungal infections increases, however, complications are becoming more prevalent[iv]. The two types of infection may instead be rapidly distinguished at point-of-care by means of a self-contained acoustic isolation and detection system, allowing doctors to immediately prescribe the appropriate remedy. The core of this system is a piezoelectric chip made from a material such as quartz or $LiNbO_3$ whose surface is functionalized with a chitin-binding lectin having a strong affinity for the cell walls of fungal pathogens. By electrically exciting this chip, Rayleigh waves lyse pathogen cells, mix the sample, and remove nonspecifically-bound materials[v], purifying chitin and isolating it on the chip's surface. The isolated analyte is then detected using a chitin-sensitive fluorescent probe. The detection limit may be further improved by means of metal-enhanced fluorescence using a thin layer of silver nanostructures formed by rapid thermal annealing that couple with the probe signal[vi]. This technique has proven effective with other analytes at concentrations on the order of picograms per milliliter[vii].

[i] McClintic, S. M., et al. "Improvement in corneal scarring following bacterial keratitis." *Eye* 27.3 (2013): 443-446.

[ii] Scott, Ingrid U., et al. "Endophthalmitis associated with microbial keratitis." *Ophthalmology* 103.11 (1996): 1864-1870.

[iii] Collier, Sarah A., et al. "Estimated burden of keratitis—United States, 2010." *MMWR. Morbidity and mortality weekly report* 63.45 (2014): 1027.

[iv] Tuberville, Audrey W., and Thomas O. Wood. "Corneal ulcers in corneal transplants." *Current eye research* 1.8 (1981): 479-485.

[v] Cular, Stefan, et al. "Removal of nonspecifically bound proteins on microarrays using surface acoustic waves." *IEEE Sensors journal* 8.3 (2008): 314-320.

[vi] Morrill, Samuel, Venkat Bhethanabotla, and Mandek Richardson. "Biomarker quantification at clinically relevant concentrations using metal enhanced fluorescence combined with surface acoustic waves." *SENSORS, 2014 IEEE*. IEEE, 2014.

[vii] Liu, Jun, Shuangming Li, and Venkat R. Bhethanabotla. "Integrating metal-enhanced fluorescence and surface acoustic waves for sensitive and rapid quantification of cancer biomarkers from real matrices." *ACS sensors* 3.1 (2018): 222-229.

606858 A Power-Free Colorimetric Biosensor for Detection of Mycobacterium Tuberculosis

Abstract:

Tuberculosis (TB) remains a significant health threat in the human society. Currently, about one-third of the human population is ill with TB worldwide due to the lack of appropriate technologies for treatment and prevention [1]. *Mycobacterium tuberculosis* is the etiological agent of tuberculosis. TB bacteria enter the body by inhalation of transmittable droplet nuclei. There are different clinical manifestations of tuberculosis such as weight loss, low-grade fever, and drenching night sweats [1].

Early diagnosis plays a major role in controlling the transmission of TB. Traditional microbial culture-based tests are the most common procedures currently used which involve cell culture, cell counts, and cell enrichment, however, this method is time consuming and difficult, particularly for the slow-growing bacteria like *M. tuberculosis*. The global public health problem of TB worldwide demands for the development of more rapid and sensitive detection methods. Currently, many methods and techniques have been developed for rapid detection of *M. tuberculosis*, such as polymerase chain reaction (PCR) [3], enzyme-linked immunosorbent assay (ELISA) [4], flow cytometry and the latest WHO-endorsed GeneXpert based test [5]. Nevertheless, they cannot provide the detection results in real-time and most of these methods are often difficult and complex to execute and centralized in large stationary laboratories because the difficulties of the operation would typically require professionals that are often expensive to retain [6]. As a result, the development of portable, real-time, sensitive, rapid, and accurate methods for *M. tuberculosis* detection is essential to effectively prevent TB infection.

Accordingly, the present study proposes a modified microchannel, a DNA nano-sensor and a portable detection system for rapid detection of mycobacterial tuberculosis enzyme activity. In the proposed detection method, a porous hydrogel was fabricated inside the microchannel via sol-gel chemistry. Polystyrene (PS) particles (45 μm) were self-assembled for 12 h under 30°C to form a closely-packed colloidal template. The poly(ethylene glycol) diacrylate solution was added dropwise on the PS template so that the solution filled the interstitial spaces within the self-assembled PS. The resulting hydrogel was cured for 30 min and then soaked in toluene for 24 h to form a porous hydrogel by removing the PS particles.

The DNA nano-sensor was prepared by conjugating a designed single DNA strand into carrier particles and AuNPs. The sensing mechanism was on the basis of cleaving hydrogen bonds between DNA strands with active MTopI and ligating the carrier particles and AuNPs. Samples containing active MTopI and DNA nano-sensor could flow along the modified microchannel driven by capillary force, at the middle of the microchannel a porous hydrogel with pore sizes smaller than the carrier particle diameter was embedded. When the MTopI was inactive, the carrier particles and AuNPs were not ligated. As a result, AuNPs were washed away from the microchannels. On the other hand, an active MTopI was capable of linking the carrier particles with AuNPs tended to aggregate in the front end of the porous hydrogel, resulting in a visible red color band in the microchannel. Readouts were performed using a smartphone-MATLAB algorithm interface. Overall, the proposed integrated platform presented in this study promotes point-of-care-testing and provides a compact and reliable tool for early diagnosis for TB patients in resource-limited areas.

- **REFERENCE**

1. WHO Global Tuberculosis Report, 2018.
2. E. Torres-Chavolla and E. C. Alocilja, "Aptasensors for detection of microbial and viral pathogens," Biosensors and Bioelectronics, vol. 24, no. 11, pp. 3175–3182, 2009.
3. D. Rodríguez-Lázaro, M. D'Agostino, A. Herrewegh, M. Pla, N. Cook, and J. Ikonomopoulos, "Real-time PCR-based methods for detection of Mycobacterium avium subsp. paratuberculosis in water and milk," International Journal of Food Microbiology, vol. 101, no. 1, pp. 93–104, 2005.
4. B. Patterson, C. Morrow, V. Singh, A. Moosa, M. Gqada, J. Woodward, V. Mizrahi, W. Bryden, C. Call, S. Patel, D. Warner, R. Wood Detection of Mycobacterium tuberculosis bacilli in bio-aerosols from untreated TB patients Gates Open Res, vol.1, pp. 11, 2017.
5. CM. Denkinger, SV. Kik, DM. Cirillo, M. Casenghi, T. Shinnick, K. Weyer, C. Gilpin, CC. Boehme, M. Schito, M. Kimerling, et al: Defining the needs for next generation assays for tuberculosis. J Infect Dis, vol. 211 Suppl 2, pp. S29-38, 2015.
6. W. ShuQi, I1. Fatih, D.L Gennaro, S. Amit, D. Utkan, Point-of-Care assays for Tuberculosis: Role of Nanotechnology/ Microfluidics. Biotechnol Adv., vol. 31(4), pp. 438–449, 2013.

612133 Evaluations of Three Low-Cost PM Monitors Against a Reference PM Instrument

Abstract:

In recent years, low-cost particulate matter (PM) monitors have significantly drawn scientific attention for their affordable cost and feasibility. These low-cost monitors are used to measure PM number and mass concentrations in various size ranges and compositions. One of the main objectives of the current study was to evaluate the accuracy of three different low-cost monitors (PupleAir, Dylos, Speck) against a reference instrument (Scanning Mobility Particle Sizer (SMPS)) in the presence of particles from grilling beef and candle burning under controlled laboratory conditions. We also evaluated the performance of the Purple Air monitor against the gravimetric mass measurement and a research level instrument, AQS1 (Aeroqual, New Zealand) in an on-campus residential apartment during several indoor activities including cooking, dishwashing, microwaving, vacuum cleaning, etc. The PurpleAir which showed good performance in the laboratory stage of the project for both candle burning and grilling beef. The correlation coefficients against SMPS were found to be 0.7 and 0.6 for grilling and candle burning, respectively. Also in the residential apartment, the average mass concentrations over 36 hours of continuous operation of Purple Air were 9.7, 14.1, and 15.0 $\mu g/m^3$ for PM_1, $PM_{2.5}$, and PM_{10}, respectively, while the PM_1, $PM_{2.5}$, and PM_{10} concentrations using the reference gravimetric measurements were 12.7, 20.8, 47.7 $\mu g/m^3$, respectively. The performance of the Purple Air was better than the AQS1 compared to the reference method in terms of the average mass concentration over 32 hours of the continuous operation. However, when a source like cooking was performed in the apartment, the Purple Air showed the peak concentration with a delay and significantly less than the research level instrument (AQS1). Time and concentration averaging as well as the method of the experiments (locating the monitors near the source or away from the source) can significantly impact the accuracy of the monitors.

598535 The Colorimetric Response of Borohydride Stabilized Silver Nanoparticle on Interaction with Organophosphates

Abstract:

Detection of residual organophosphates in food is of paramount importance in the current scenario. Biosensors for pesticide detection may be developed by monitoring the interaction of nanoparticles with pesticides. In this report, we have studied the colorimetric response of pesticides phorate, chlorpyrifos and malathion with borohydride stabilized silver nanoparticle. The colorimetric changes are corroborated with UV-Visible spectra along with the change in particle size and zeta potential. Nanoparticle on interaction with phorate show visible change in color to dark green while other two pesticides haven't shown any visible color change. The colorimetric response of nanoparticle pesticide mixture was further studied for 12 hours of incubation and it was observed that nanoparticle-phorate mixture started decolorizing after an hour and becomes colorless after 12 hours of incubation, while other two pesticide mixtures remains same. This decolorization of phorate was correlated with the degradation of pesticide in the presence of bare silver nanoparticle. The reaction mechanism was explained by the formation of silver and sulfur atom (Ag→S) complex. The uncapped nanoparticles get stabilized by pesticides over time for chlorpyrifos and malathion while phorate undergoes degradation due to scission of $S\text{-}CH_2$ present in the molecule. This color change may be utilized as a colorimetric method for the detection of phorate pesticide.

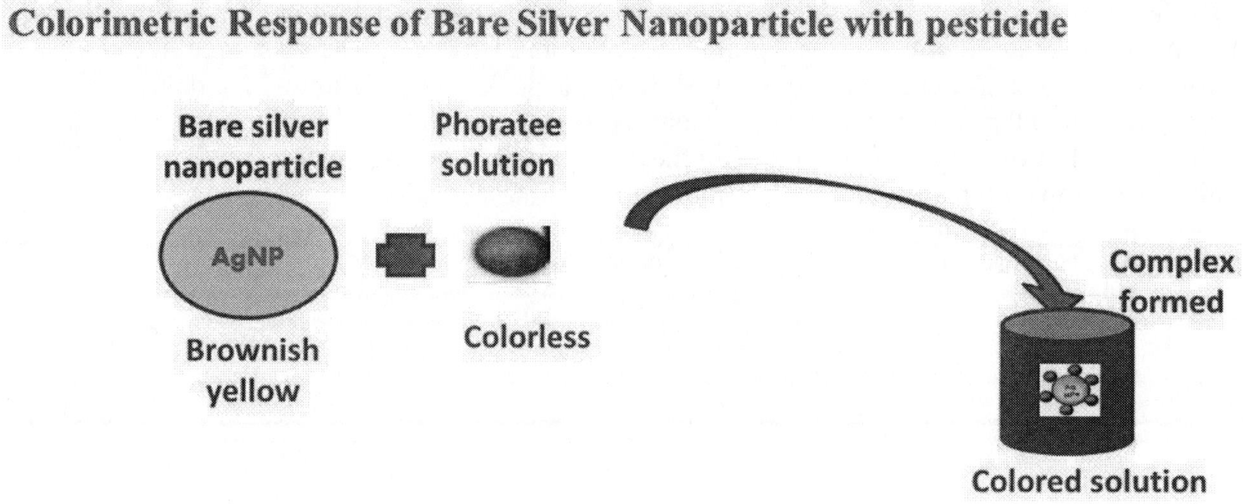

598537 Analysis of Organophosphate Residue Detection on SOLID Surface Using QCM-D As Sensing Platform

Abstract:

Detection of residual organophosphates in food is of paramount importance in current scenario. The understanding of interaction of pesticides on chemically modified surface is the backbone of biosensor fabrication. Quartz crystal microbalance with dissipation is one of the most efficient techniques to analyze the interaction of organic molecules with silane modified surfaces making it desirable bio-sensing platform.

In this report, we have studied the real time adsorption of pesticide phorate and malathion on silica coated quartz crystal of QCM-D. The interaction study was first monitored on gold coated and unmodified silica coated crystal as reference. Further self-assembled monolayer was created on silica crystal using n-propyltrimethoxy silane (PTMS), octyltrimethoxy silane (OTMS), N-(3-(trimethoxysilyl) propyl) aniline and diethoxymethylphenyl silane. The interaction study was also monitored by using AFM analysis. Amount of pesticide adsorbed on surface was evaluated in terms of the mass and thickness change of quartz crystal which is calculated from frequency and dissipation factor change using Sauerbrey equation. Low value of dissipation factor (ΔD) corresponds to stronger binding of pesticide on crystal. The overall results indicates that OTMS modified surfaces contributes mostly to the dissipation changes (ΔD) or acoustic ratio ($\Delta D/\Delta f$). This surface shows largest amount of pesticide adsorption for both malathion and phorate which is correlated with the more hydrophobicity of the surface analyzed by contact angle measurements. The adsorbed amount of pesticide increases with increase in hydrophobicity of surface. The diethoxymethylphenyl silane also shows significant interaction as the phenyl group facilitates interaction with sulfur bond of pesticide. Further verification was done using Raman spectroscopy. The interaction study may be utilized further to fabricate sensors for pesticide detection.

603653 Energy Harvesting Wireless Sensors Using Magnetic Phase Transition

Abstract:

Most industries are connected by cyber-physical systems (CPS), and data and energy in manufacturing technologies are exchanged and shared or sustainable development. To deploy CPS, it is necessary to develop overall security systems, to increase their efficiency including data transfer systems in the network and acquisition systems, and to find new energy sources.

To overcome these energy-related issues of CPS, the possibility of integrating data acquisition sensors with energy harvesting system has been investigated to design energy harvesting sensors in this research. In this energy harvesting system, the magnetic phase transition resulting from changes in temperature is combined with electromagnetic induction, leading to conversion from heat to electricity in the system. In addition, we can successively measure the electromotive force as sensing targets. The proposed system can provide wireless temperature or velocity sensors that directly measure electromotive forces generated by Faraday's law of induction without any additional energy input. Our proposed energy harvesting sensors have the promising potential to contribute significantly to the development of CPS in the near future.

608134 Ultrastretchable Conductive Polymer Complex As a Strain Sensor with a Repeatable Autonomous Self-Healing Ability

Abstract:

Wearable strain sensors are essential for the realization of applications in the broad fields of remote healthcare monitoring, soft robots, and immersive gaming, among many others. These flexible sensors should be comfortably adhered to the skin and capable of monitoring human motions with high accuracy, as well as exhibiting excellent durability. However, it is challenging to develop electronic materials that possess the properties of skin - compliant, elastic, stretchable, and self-healable. This work demonstrates a new regenerative polymer complex composed of poly(2-acrylamido-2-methyl-1-propanesulfonic acid), polyaniline, and phytic acid as a skin-like electronic material. It exhibits ultrahigh stretchability (1935%), repeatable autonomous self-healing ability (repeating healing efficiency >98%), quadratic response to strain ($R^2 > 0.9998$), and linear response to flexion bending ($R^2 > 0.9994$), outperforming current reported wearable strain sensors. The deprotonated polyelectrolyte, multivalent anion, and doped conductive polymer, under ambient conditions, synergistically construct a regenerative dynamic network of polymer complex cross-linked by hydrogen bonds and electrostatic interactions, which enables ultrahigh stretchability and repeatable self-healing. Sensitive strain responsive geometric and piezoresistive mechanisms of the material owing to the homogeneous and viscoelastic nature provide excellent linear responses to omnidirectional tensile strain and bending deformations. Furthermore, this material is scalable and simple to process in an environmentally friendly manner, paving the way for the next-generation flexible electronics.

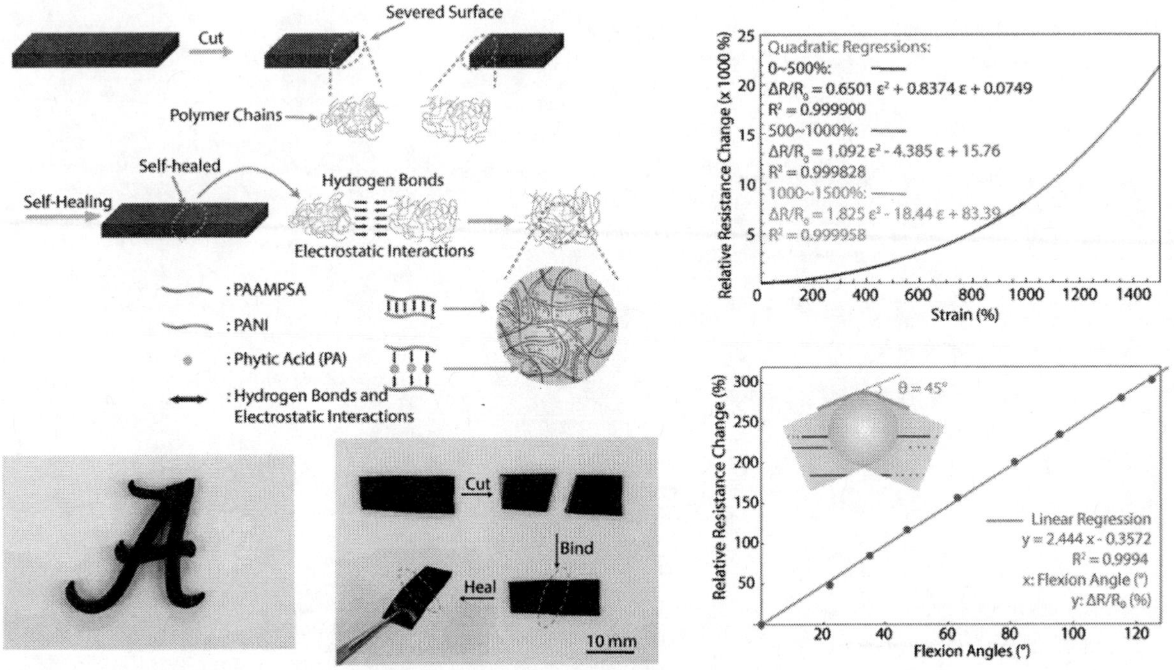

610602 Monitoring Inconsistent ELP Surface Modification By Thiol Desorption

Abstract:

Elastin-like polymers (ELPs) have gained attention recently for their potential as a transduction enhancement for surfaces. However, consistent surface modification is required to ensure a consistent response model. Gold surfaces with ELP has been a challenge in practice, but can be achieved with careful modification procedure, keeping variables such as temperature, ELP concentration, time of modification, and dissolution procedure consistent. We hypothesize that further ELP surface modification inconsistencies are due to ELP that physiosorbs to the gold surface instead of forming new chemisorption modification sites. This presentation will share data demonstrating that inconsistencies in electrochemical measurements of surface-bound ELP are due to inconsistent physisorption of non-chemisorbed ELP molecules. Cyclic voltammetry in KOH was applied to surface-bound ELP at different ELP concentrations in order to reduce the ELP - gold thiol surface. The area under the reduction peak is equal to amount of ELP directly bound to the surface. Electrochemical impedance spectroscopy (EIS) was used with a Randles circuit model to determine the charge transfer resistance of the surfaces. By comparing the trend of charge transfer resistance to surface-bound ELP, we show that optimization coverage points for ELP modified surfaces.

597825 CRISPR-Cas Systems Mediated Electrochemical Biosensing Platforms

Abstract:

Robust developments of personalized medicine for next-generation healthcare highlight the need for sensitive and accurate point-of-care platforms for quantification of disease biomarkers. Broad presentations of clustered regularly interspaced short palindromic repeats (CRISPR) as an accurate gene editing tool indicate that the high-specificity and programmability of CRISPR system can be utilized as a powerful biosensing tool. Herein, we introduce how CRISPR-Cas systems can modulate electrochemical biosensing platforms toward sensitivity, accuracy and programmability. The presented examples demonstrate the potential of CRISPR-Cas systems based biosensing platforms for the development of ideal integrated point-of-care devices.

602395 Dynamic Data Reconciliation, Parameter Estimation, and Validation for the Dynamic Model of a Supercritical Pulverized Coal Power Plant

Abstract:

The need for process systems engineering tools to improve the performance of existing fossil-fueled power plants in the presence of renewables in the energy market is well documented [1]. A powerful, low risk, and cost-effective tool for the improvement of these existing systems is the development of highly predictive process models. In the past, many such models for coal-fired power plants have been developed [2], [3]. However, many of these models focus on the steady-state design and operation of traditional subcritical pulverized coal systems, with less consideration given to dynamic models of supercritical pulverized coal (SCPC) power plant. A few papers are focused on the impact of carbon capture on the SCPC plant [4]–[6] where simple models of the SCPC plant were used. One fairly comprehensive study has been done using the commercial software Apros, but the equipment models used lack some of the fidelity needed to capture the dynamics of the system [7]. For a general SCPC plant, a past contribution established a plantwide model with higher fidelity for some systems in the plant [8]. A particular gap in the literature is the lack of industrial validation for the models that have been developed.

In this presentation, a first-principles comprehensive dynamic model of a SCPC plant is described comprising detailed models of the following sections - boiler, turbine, feed water heaters, and the balance of the plant. Distributed models for each component of the boiler including the economizer, superheaters, water wall, and platen superheater are discussed. Each section of the boiler model also contains a detailed heat transfer model with due consideration of conductive/convective/radiative heat transfer as appropriate. The boiler model takes into account the changes in the heat transfer coefficients due to the hydrodynamics of the system and thermal properties of the steam that can vary rapidly during load-following especially under sliding pressure operation. The model of the steam turbine can capture the change in the performance characteristics of the leading, non-condensing, and condensing stages. In addition, the steam turbine model can detect and account for moisture, if present, in the stages that may be non-condensing under nominal condition.

The SCPC dynamic model is validated with the full-load and transient data from an industrial partner plant (IPP). The data does not satisfy material and energy balances motivating the solution of a dynamic data reconciliation problem to ensure closure of the balances [9], [10]. One difficulty in the dynamic data reconciliation for this system is the poor estimate of the holdup - especially thermal holdup in the boiler components - and the poor initial guess of the initial state of the system. It is also impractical to mimic the complex feed-forward-augmented feedback controllers and logic-based approaches used in the coordinated control system in the IPP. Furthermore, there are large number of model parameters, especially equipment specific parameters that are not available and must be estimated. Due to the large number of partial differential algebraic equations in the plant-wide model, the optimization problem for dynamic data reconciliation and parameter estimation is large. A two-stage optimization problem is solved for reconciling the dynamic data in the first stage followed by optimal estimation of the parameters. Uncertainties in the initial estimates of the material and thermal holdup are incorporated in the form of sensitivity information in the proposed approach for reducing the computational expense. This work shows that steady-state and transient performance of the IPP can be very well predicted by the model. The validated SCPC dynamic model is used for developing advanced model-based controllers for load-following operation. The controllers maximize the plant efficiency during transient operation while satisfying temperature constraints at various locations. Results are also shown for important calculated performance indices, such as the plant heat rate.

[1] A. Subramanian, T. Gundersen, and T. A. Adams II, "Modeling and Simulation of Energy Systems: A Review," *Processes*, vol. 6, p. 238, 2018.

[2] E. Oko and M. Wang, "Dynamic modelling, validation and analysis of coal-fired subcritical power plant," *Fuel*, vol. 135, pp. 292–300, Nov. 2014, doi: 10.1016/j.fuel.2014.06.055.

[3] C. Chen, Z. Zhou, and G. M. Bollas, "Dynamic modeling, simulation and optimization of a subcritical steam power plant. Part I: Plant model and regulatory control," *Energy Convers. Manag.*, vol. 145, pp. 324–334, Aug. 2017, doi: 10.1016/j.enconman.2017.04.078.

[4] A. K. Olaleye, E. Oko, M. Wang, and G. Kelsall, "Dynamic Modelling and Analysis of Supercritical Coal-Fired Power Plant Integrated with Post-combustion CO2 Capture," in *Clean Coal Technology and Sustainable Development*, Singapore, 2016, pp. 359–363.

[5] K. Wellner, T. Marx-Schubach, and G. Schmitz, "Dynamic Behavior of Coal-Fired Power Plants with Postcombustion CO2 Capture," *Ind. Eng. Chem. Res.*, vol. 55, no. 46, pp. 12038–12045, Nov. 2016, doi: 10.1021/acs.iecr.6b02752.

[6] S. Ó. Garđarsdóttir, R. M. Montañés, F. Normann, L. O. Nord, and F. Johnsson, "Effects of CO2-Absorption Control Strategies on the Dynamic Performance of a Supercritical Pulverized-Coal-Fired Power Plant," *Ind. Eng. Chem. Res.*, vol. 56, no. 15, pp. 4415–4430, Apr. 2017, doi: 10.1021/acs.iecr.6b04928.

[7] J. Hentschel, H. Zindler, and H. Spliethoff, "Modelling and transient simulation of a supercritical coal-fired power plant: Dynamic response to extended secondary control power output," *Energy*, vol. 137, pp. 927–940, Oct. 2017, doi: 10.1016/j.energy.2017.02.165.

[8] P. Sarda, E. Hedrick, K. Reynolds, D. Bhattacharyya, E. S. Zitney, and B. Omell, "Development of a Dynamic Model and Control System for Load-Following Studies of Supercritical Pulverized Coal Power Plants," *Processes*, vol. 6, no. 11, 2018, doi: 10.3390/pr6110226.

[9] A. S. Chinen, J. C. Morgan, B. Omell, D. Bhattacharyya, and D. C. Miller, "Dynamic Data Reconciliation and Validation of a Dynamic Model for Solvent-Based CO2 Capture Using Pilot-Plant Data," *Ind. Eng. Chem. Res.*, vol. 58, no. 5, pp. 1978–1993, Feb. 2019, doi: 10.1021/acs.iecr.8b04489.

[10] P. Mobed, J. Maddala, R. Rengaswamy, D. Bhattacharyya, and R. Turton, "Data Reconciliation and Dynamic Modeling of a Sour Water Gas Shift Reactor," *Ind. Eng. Chem. Res.*, vol. 53, no. 51, pp. 19855–19869, Dec. 2014, doi: 10.1021/ie500739h.

605798 Forecasting Prices of Energy Feedstocks and Commodities Using Advanced Statistical and Machine Learning Methods

Abstract:

The prices of energy products and commodities are volatile [1] and sensitive to shifts in demand and supply mechanisms, technological breakthroughs, changes in monetary policy, geopolitical events or other major global challenges [2, 3]. Considering the fact that energy affects every single individual and entity in the world, it is crucial to quantify "the price of energy" accurately, and study how it evolves over time. To this respect, we have developed a novel forecasting framework to calculate the average as well as the market (spot) price of energy in the United States by introducing two indices: The Energy Price Index (EPIC) and the Energy Spot Price Index (ESPIC) respectively [4].

The complex energy landscape is thoroughly analyzed [5-9] to accurately determine the two key factors of this framework: the total demand of the energy products directed to the end-use sectors, and the corresponding price of each product in the form of either a monthly or a spot price [10-14]. We have already presented a rolling horizon forecasting methodology for the demands of energy in the future [4], and here we demonstrate a novel technique to forecast the future prices of energy feedstocks and commodities up to 12 months. The historical prices of each energy product and commodity are studied individually, so as to identify patterns, trends, cycles, outliers, relationships among different variables etc. Then, different forecasting models are fitted in training data sets i.e. Exponential Smoothing, ARIMA, Dynamic Regression, Neural Networks etc. The best forecasting model for each energy product and commodity is selected based on the validation results from testing data sets, and is then utilized to produce the 12 months forecasts. Being able to forecast the future prices of energy has tremendous potential for applications in economics, finance, engineering, law and policy.

References:

1. Regnier, E., 2007. Oil and energy price volatility. *Energy economics*, *29*(3), pp.405-427
2. I.W. Parry, M. D. Heine, E. Lis, and S. Li, Getting energy prices right: From principle to practice. International Monetary Fund, 2014.
3. Barnett, J.; Barron, J.; Oil market volatility is at all-time high. Energy Information Administration (EIA), Today in Energy, March 2020.
4. Baratsas, S.; Niziolek, A.; Onel, O.; Matthews, L.; Floudas, C. A.; Hallermann, D.; Sorescu, S.; Pistikopoulos, E. N. What is the price of energy to the end-users? A novel predictive framework and one of its applications to the energy and monetary policies. Nature Communications 2020 (Under Review).
5. Elia, J. A.; Baliban, R. C.; Xiao, X.; Floudas, C. A. Optimal energy supply network determination and life cycle analysis for hybrid coal, biomass and natural gas to liquid (CBGTL) plants using carbon-based hydrogen production. Computers and Chemical Engineering 2011, 35, 1399-1430.
6. Elia, J. A.; Baliban, R. C.; Floudas, C. A. Nationwide energy supply chain analysis for hybrid feedstock processes with significant CO2 emissions reduction. AIChE Journal 2012, 58 (7), 2142-2154.
7. Elia, J. A.; Floudas, C. A. Energy Supply Chain Optimization of Hybrid Feedstock Processes: A Review. Annual Review of Chemical and Biomolecular Engineering 2014, 5 (1), 147-179.

8. Onel, O.; Niziolek, A. M.; Floudas, C. A. Integrated biomass and fossil fuel systems towards the production of fuels and chemicals: state of the art approaches and future challenges. Current Opinion in Chemical Engineering 2015, 9, 66-74.
9. Floudas, C.A.; Niziolek, A. M.; Onel, O.; Matthews, L.R. Multi-Scale Systems Engineering for Energy and the Environment: Challenges and Opportunities. AIChE Journal 2016, 62 (3), 602-623.
10. Energy Information Administration (EIA), Annual Energy Outlook, 2019
11. Energy Information Administration (EIA), Monthly Energy Review, 2019
12. Lazard, Levelized Cost of Energy Analysis Version 13.0, 2019
13. Bureau of Labor Statistics (BLS), Producers Price Index, 2019
14. S. Department of Energy (DOE), Alternative Fuels Data Center, 2019

602323 Implementation of Hybrid Models to Perform System Analyses with Model Maintenance in Continuous Pharmaceutical Manufacturing

Abstract:

Recent development of Industry 4.0 technologies, including the advancements of cyber-physical systems (CPS) and Internet of Things (IoT), have increased the amount of data available from manufacturing shop floors and analytical laboratories[1-5]. The abundance of data contributes to the emergence and improvement of data analytic methods and cloud computing platforms, aiming to extract more information and knowledge from data to support predictive modeling and decision making[6, 7]. Traditionally, process industries, including continuous pharmaceutical manufacturing (CPM), mainly use white-box (WB) models for predictive modeling[8]. These WB models, including mechanistic, phenomenological, and first-principle models, rely heavily or solely on physical understandings of the system. Though the WB models provide a high degree of transparency on addressing the physical significance of variables and model structures, the development of the models is costly, and the prediction performance is largely dependent on the mastery of full process knowledge. To address these challenges utilizing the vast availability of data, recent efforts have focused on combining data with process knowledge to form hybrid models (HMs). HMs consist of at least one white-box and one black-box (BB) sub-model, where the BB sub-models take in historical process data and adopt different machine learning (ML) algorithms like artificial neural network (ANN) and support vector machine (SVM)[9]. The development of HMs of various structures provides greater flexibility and better predictability in complex engineering systems, making it applicable in modeling CPM processes.

Current research has been dedicated to developing proof-in-concept HMs for individual unit operations of CPM processes, but the capability of HMs to perform system analyses like sensitivity and feasibility analysis remains unexplored. These analyses can identify critical process parameters and design space of CPM systems, which is crucial for process design and operations[10]. Another challenge of implementing HMs in CPM systems is related to model maintenance. Parts of HMs are data-driven BB sub-models, and the offline training of BB sub-models with defined sets of historical data leads to static BB models[11]. These static BB sub-models only reflect the system at the timepoint that the models are developed, and they will not change unless the models are re-trained with new data and algorithms[12]. As time progresses, assuming the process stays the same, new datasets collected from plants should have the same or similar underlying distribution as the data used to train the original model. Using these new datasets to re-train should yield models with comparable performance to their predecessors. However, real datasets from manufacturing plants often contain normal operational data and abnormal data (e.g. highly noisy data, faulty data, missing data points, testing and maintenance data), and the quality of these new data can affect the re-training result considerably. On the other hand, like in the majority of manufacturing processes, a shift in the distributions of manufacturing data from CPM plants can happen over time due to changes such as environmental conditions, equipment performance, and control strategy. Because these changes are not captured in the model training process, deteriorating performance of HMs can be observed, which is a classic scenario known as model drift[11]. Model maintenance strategies therefore need to be in place to capture this shift and to ensure the accuracy of model predictions.

In this work, algorithms are developed to perform system analyses using HMs and to address model maintenance issues in CPM. For system analyses, HMs developed for individual unit operations of the continuous direct compaction (DC) line are connected in a flowsheet model, and they are then used to perform sensitivity and feasibility analysis[10, 13, 14]. The results from HMs are compared to those produced by WB models to compare the model performance. Next, two tasks are carried out to explore model maintenance schemes for HMs. Different types and amounts of new datasets are first fed into the re-training process to study the impact of the quality of additional data to HM performance. This analysis can yield practical guidance in data handling and maintenance frequency based on characteristics of newly available datasets. Then, to address the issue of model drift, different learning techniques and adaptive methods for model maintenance are developed. Manual re-training, incremental learning, and continuous streaming procedures are explored along with blind and informed adaptive algorithms[11, 15, 16]. Blind adaptations update an HM continuously without realizing that a drift exists, whereas informed algorithms only become effective when predefined triggers are activated[11]. The results will provide insights on how model maintenance for HMs can be performed. Together with system analysis capabilities, this will allow for broader acceptance of HMs for practical use in CPM with the development of Industry 4.0 framework.

1. Ding, B., Pharma Industry 4.0: Literature review and research opportunities in sustainable pharmaceutical supply chains. *Process Safety and Environmental Protection* 2018, *119*, 115-130.
2. Cao, H.; Mushnoori, S.; Higgins, B.; Kollipara, C.; Fermier, A.; Hausner, D.; Jha, S.; Singh, R.; Ierapetritou, M.; Ramachandran, R., A Systematic Framework for Data Management and Integration in a Continuous Pharmaceutical Manufacturing Processing Line. *Processes* 2018, *6* (5).
3. Reis, M. S.; Kenett, R., Assessing the value of information of data-centric activities in the chemical processing industry 4.0. *AIChE Journal* 2018, *64* (11), 3868-3881.
4. Barenji, R. V.; Akdag, Y.; Yet, B.; Oner, L., Cyber-physical-based PAT (CPbPAT) framework for Pharma 4.0. *Int J Pharm* 2019, *567*, 118445.
5. Steinwandter, V.; Borchert, D.; Herwig, C., Data science tools and applications on the way to Pharma 4.0. *Drug Discov Today* 2019, *24* (9), 1795-1805.
6. Thomas, M. C.; Zhu, W.; Romagnoli, J. A., Data mining and clustering in chemical process databases for monitoring and knowledge discovery. *Journal of Process Control* 2018, *67*, 160-175.
7. Liu, Y.; Xu, X., Industry 4.0 and Cloud Manufacturing: A Comparative Analysis. *Journal of Manufacturing Science and Engineering* 2016, *139* (3).
8. Zendehboudi, S.; Rezaei, N.; Lohi, A., Applications of hybrid models in chemical, petroleum, and energy systems: A systematic review. *Applied Energy* 2018, *228*, 2539-2566.
9. von Stosch, M.; Oliveira, R.; Peres, J.; Feyo de Azevedo, S., Hybrid semi-parametric modeling in process systems engineering: Past, present and future. *Comput Chem Eng* 2014, *60*, 86-101.
10. Wang, Z.; Escotet-Espinoza, M. S.; Ierapetritou, M., Process analysis and optimization of continuous pharmaceutical manufacturing using flowsheet models. *Comput Chem Eng* 2017, *107*, 77-91.
11. Gama, J.; Žliobaitė, I.; Bifet, A.; Pechenizkiy, M.; Bouchachia, A., A survey on concept drift adaptation. *ACM Computing Surveys* 2014, *46* (4), 1-37.
12. Webb, G. I.; Hyde, R.; Cao, H.; Nguyen, H. L.; Petitjean, F., Characterizing concept drift. *Data Mining and Knowledge Discovery* 2016, *30* (4), 964-994.
13. Metta, N.; Ghijs, M.; Schäfer, E.; Kumar, A.; Cappuyns, P.; Assche, I. V.; Singh, R.; Ramachandran, R.; Beer, T. D.; Ierapetritou, M.; Nopens, I., Dynamic Flowsheet Model Development and Sensitivity Analysis of a Continuous Pharmaceutical Tablet Manufacturing Process Using the Wet Granulation Route. *Processes* 2019, *7* (4).

14. Boukouvala, F.; Niotis, V.; Ramachandran, R.; Muzzio, F. J.; Ierapetritou, M. G., An integrated approach for dynamic flowsheet modeling and sensitivity analysis of a continuous tablet manufacturing process. *Comput Chem Eng* 2012, *42*, 30-47.
15. Kadlec, P.; Grbić, R.; Gabrys, B., Review of adaptation mechanisms for data-driven soft sensors. *Comput Chem Eng* 2011, *35* (1), 1-24.
16. Pechenizkiy, M.; Bakker, J.; Žliobaitė, I.; Ivannikov, A.; Kärkkäinen, T., Online mass flow prediction in CFB boilers with explicit detection of sudden concept drift. *ACM SIGKDD Explorations Newsletter* 2010, *11* (2), 109-116.

593037 Utilization of Advanced Analytics to Monitor Catalyst Health in an Ethylene Oxide Reactor

Abstract:

Ethylene oxide (EO), a precursor to ethylene glycol (EG), is produced through elective oxidation of ethylene and oxygen in the presence of an Ag supported α-Al2O3 catalyst in the EO/EG plant located in the Petkim Petrochemical Holding complex in Aliaga, Turkey. The EO process takes place in two parallel, fixed-bed multi-tubular reactors at a temperature and pressure range of 240–260 °C and 17–18 bar, respectively. Both ethylene epoxidation and EO combustion take place in the reactors and only the former is desirable. Side reactions result in a reduction of EO yield and an increase in CO2 emission from the plant. Thus, the optimization of the catalyst exploitation continuance becomes an important concern to maximize the primary reaction for product quality and plant profitability.

In this presentation, machine learning methods are utilized in the characterization and monitoring of the industrial catalyst used in the EO reactors. Estimation of the catalyst activity and other indicators has major importance during production due to the five-year lifetime with high catalyst cost and the EO product quality. In order to make accurate estimates, a vast quantity of data is collected from the plant and is processed using several statistical methods in order to capture the major driving signals that provide significant implications. Principal component analysis (PCA) is used for dimensional reduction and to categorize the measurements. The performances of various machine learning algorithms are compared for the prediction of that catalyst status. The machine learning model provides a good insight into the causes of rapid catalyst deactivation. Once these insights are obtained, the operating policies and process conditions that will maximize the catalyst lifetime and other desirable properties of the catalyst in the plant will be discussed in detail.

600749 Collaboration Is Critical for Effective Deployment of Big Data Analytics

Abstract:

Manufacturers across the world have begun their digital transformations with an eye towards utilizing big data analytics. With this effort, process sensors are providing a wealth of data, numerous machine learning algorithms are being developed, and data scientists are being hired at an astounding rate. While these advancements are important, the reality is that a wealth of key information, or context, is typically not captured in any database. That wealth of information is the internal expertise that subject matter experts, like engineers and scientists, rely on to drive effective decision making. Considering this, an effective analytics strategy must utilize technology and workflows that combine both human experience and computer technology.

In collaboration, subject matter experts are relied upon throughout the project to provide relevant context, document assumptions, and explain applicable constraints. Data scientists complement this information with their knowledge of machine learning algorithms and other relevant analytics options. Together, the team finds the hidden patterns in the data that translate to business value for the company.

This session will provide a framework for leveraging analytics applications to promote effective collaboration between subject matter experts and data scientists to accelerate the time to insight and operationalize models and algorithms. The presentation will cover the importance of subject matter expert involvement in data preparation for successful deployment of models and algorithms through a series of examples of effective analytics deployment. Examples include operations selected from a variety of industries, including pharmaceuticals and chemicals.

608672 Dynamic Data Feature Engineering for Process Operation Troubleshooting

Abstract:

Data analytics and machine learning have shown tremendous power in mining and analyzing a large amount of data to reveal useful information and knowledge behind the data. Many industries, including chemicals, petrochemicals, energy, power grids, and pharmaceutical, have turned their attention to the massive data they own for improvement in efficiency and competitiveness. Owing to rich instrumentation and control in engineering and manufacturing systems and the deployment of industrial internet of things (IIoT), data collected from industrial operations are of high dimensions and fast sampling. Therefore, there is a great need to develop analytics tools geared towards analyzing massive process data.

Most process data are collected in the form of time series, which are highly cross-correlated and auto-correlated. In other words, the high-dimensional data are not dynamically excited in all dimensions of the measurement space. This situation is more pronounced with IIoT where sensors are installed with a degree of redundancy which leads to collinearity. Due to collinearity, the dynamic variations in plant data are often concentrated in a low dimensional subspace. On the other hand, the complementary subspace has only random variations which are independent over time. Because of these characteristics, traditional modeling tools such as vector autoregressive moving average (VARMA) analysis are not suitable since they assume full dimensional dynamics (Tsay, 2013).

In this paper we apply the dynamic inner canonical correlation analysis (DiCCA) developed in Dong and Qin (2018) and Dong et al. (2020) to extract low dimensional latent variables. Each of the latent variable models is a self-dependent univariate autoregressive (AR) model. The latent variables are orthogonal or contemporaneously independent of each other, which is convenient for visualizing latent features and troubleshooting abnormal variations in high dimensional data. In addition, each of the latent variables is rank-ordered by the predictability from its own history. This objective promotes self-dependent AR relations to be extracted. For example, integrating components and oscillatory components are favored by this objective.

In this paper, we propose a dynamically engineered latent feature analysis (DELFA) procedure for plant-wide troubleshooting by applying the DiCCA algorithm to decompose high dimensional process data into dynamic latent features. DELFA does not make use of the prediction model of DiCCA. It finds dynamic features of a segment of time series data that contain interesting features, which could be associated with anomalies. We also extend the DiCCA algorithm to deal with exogenous variables, which is referred to the DiCCAX algorithm.

DELFA further identifies measured variables that are best interpreted by the latent features.

The degree of interpretation is represented by the latent variable loadings. Composite loadings and weights are derived to analyze features that appear in multiple latent variables.

The features of interest can be intermittent in time; when they happen their loadings on measured variables are the focus of analysis. We demonstrate the effectiveness of the DELFA troubleshooting procedure on two high dimensional datasets from an industrial plant. One dataset is analyzed with the troubleshooting procedure to find several anomalous features. The other data set, collected after fixing a major anomaly, is analyzed to confirm the fix of the major anomaly and also find other anomalies.

References

Ruey S. Tsay. Multivariate Time Series Analysis: with R and financial applications. John Wiley & Sons, 2013.

Yining Dong and S Joe Qin. Dynamic latent variable analytics for process operations and control. Computers & Chemical Engineering, 114:69-80, 2018.

Yining Dong, Yingxiang Liu, and S Joe Qin. Efficient dynamic latent variable analysis for high dimensional time series data. IEEE Trans. on Industrial Informatics, vol. 16, no. 6, pp. 4068-4076, June 2020.

614025 (Invited Talk) Counting Molecules, Dodging Blood Cells: Continuous, Real-Time Molecular Measurements Directly in the Living Body

Abstract:

The availability of technologies capable of tracking the levels of drugs, metabolites, and biomarkers in real time in the living body would revolutionize our understanding of health and our ability to detect and treat disease. Imagine, for example, a dosing regime that, rather than relying on your watch ("take two pills twice a day"), is instead guided by second-to-second measurements of plasma drug levels wirelessly communicated to your smartphone. Such a technology would likewise provide researchers and clinicians an unprecedented window into neurology and physiology, and could even support ultra-high-precision personalized medicine in which drug dosing is optimized minute-by-minute using closed-loop feedback control. Towards this goal, we have developed a biomimetic,electrochemical sensing platform that supports the high frequency, real-time measurement of specific molecules (irrespective of their chemical reactivity) in situ in the blood and tissues of awake, freely moving subjects. Key to achieving this was the development of sensors small enough to emplace in the body but retaining sufficient s/n in such a complex environment. This was done via the use of nanostructured and nanoporous electrode materials of greatly enhanced microscopic area.

610717 Ultrasensitive, Selective, and Reversible Room-Temperature NO_2 Sensor Based on a Monolayer Transition Metal Dichalcogenide

Abstract:

Chemical sensors are essential for environmental and health monitoring. In particular, gas sensors that can efficiently detect toxic gases emitted from power plants, vehicles, and industrial sources at room temperature, are of substantial importance. Here, we demonstrate a room-temperature NO_2 sensor based on a monolayer transition metal dichalcogenide that is highly sensitive and selective. Depending on the layer number of the sensing devices, their electrical resistances substantially increase or decrease upon exposure to NO_2. We find the atomically-thin sensors to be strongly selective to NO_2 with only minimal responses to NH_3, CH_2O, and CO_2. Importantly, the monolayer sensor shows complete reversibility with fast recovery in ambient condition (*i.e.* at room temperature and under substantial relative humidity). Combined with theoretical calculations, we explore interactions between different molecular species and the monolayer crystal and discuss the underlying sensing mechanism.

605192 Atomically-Thin Sensing Surfaces from 2D Materials for Detecting Cellular Gaps

Abstract:

Cellular layers play critical role as barriers, filters, and membranes in our bodies. The performance of cellular layers is often found to be determined by the properties of individual cellular gaps. Yet these gaps often fall below the diffraction limit, hampering their detection. Here, we develop a novel sensing modality called chemical tomography, where the diffusion of a harmless chemical tracer is used to reveal cellular gaps in human umbilical vein endothelial cellular (HUVEC) layers. For the readout, we employ atomically-thin and optically-active MoS2, whose two-dimensional form-factor enables it to serve as an array of 20,164 wireless sensors over 420x420 μm^2 area. A rigorous screening of 30 analytes identifies several tracer candidates with MoS2 dynamic range over five orders of magnitude in tracer concentration. In a single layer of HUVEC cells, we find local spots with 10^{-8}-10^{-9} cm^2/s diffusivities that are spatially correlated with immunofluorescence of vascular endothelial (VE)-cadherin proteins found on cell membrane. These spots correspond to cellular pores with effective sizes from 50 to 100 nm. This platform enables future studies on properties and dynamics of cellular gaps of various tissues.

607027 Nanocomposites of Multiwalled Carbon Nanotubes and Palladium-Based Nanosheets for Hydrogen Sensing

Abstract:

Thin palladium nanosheets have been gaining attention for various applications due to their high surface area, high surface atom fraction, and excellent electrical conductivity. Palladium has been explored as a resistive hydrogen sensor because of its unique property of forming hydrides in the presence of hydrogen at room temperature. However, most Pd-based sensors suffer from low response with a limit of detection around 300 ppm. This work describes the design and fabrication of multi-walled carbon nanotubes@Pd nanosheets (MWCNT@Pd NS) composites for hydrogen sensing applications. The Pd nanosheets were prepared by a CO confinement strategy using a one-pot solvothermal method. The prepared Pd nanosheets had a high surface area of ~125 m^2/g with a thickness of only ~2 nm. The prepared Pd nanosheets were then dispersed in hexane along with CNT and sonicated to produce a homogeneous suspension. The nanocomposite solution was drop cast onto an interdigitated electrode, and its performance for hydrogen detection in synthetic air was evaluated. Compared with pristine Pd nanosheets, the MWCNT@Pd nanosheets showed enhanced hydrogen detection performance. The resistance change of MWCNT@Pd nanosheets to 2% H_2 was 2%, compared to 0.5% change for pristine Pd nanosheets. The LOD of MWCNT@Pd nanosheets was also found to improve to <10 ppm, far below the 300 ppm LOD of Pd nanosheets. This encouraging result could prove useful in the fabrication of low-cost hydrogen sensors based on 2D Pd nanomaterials.

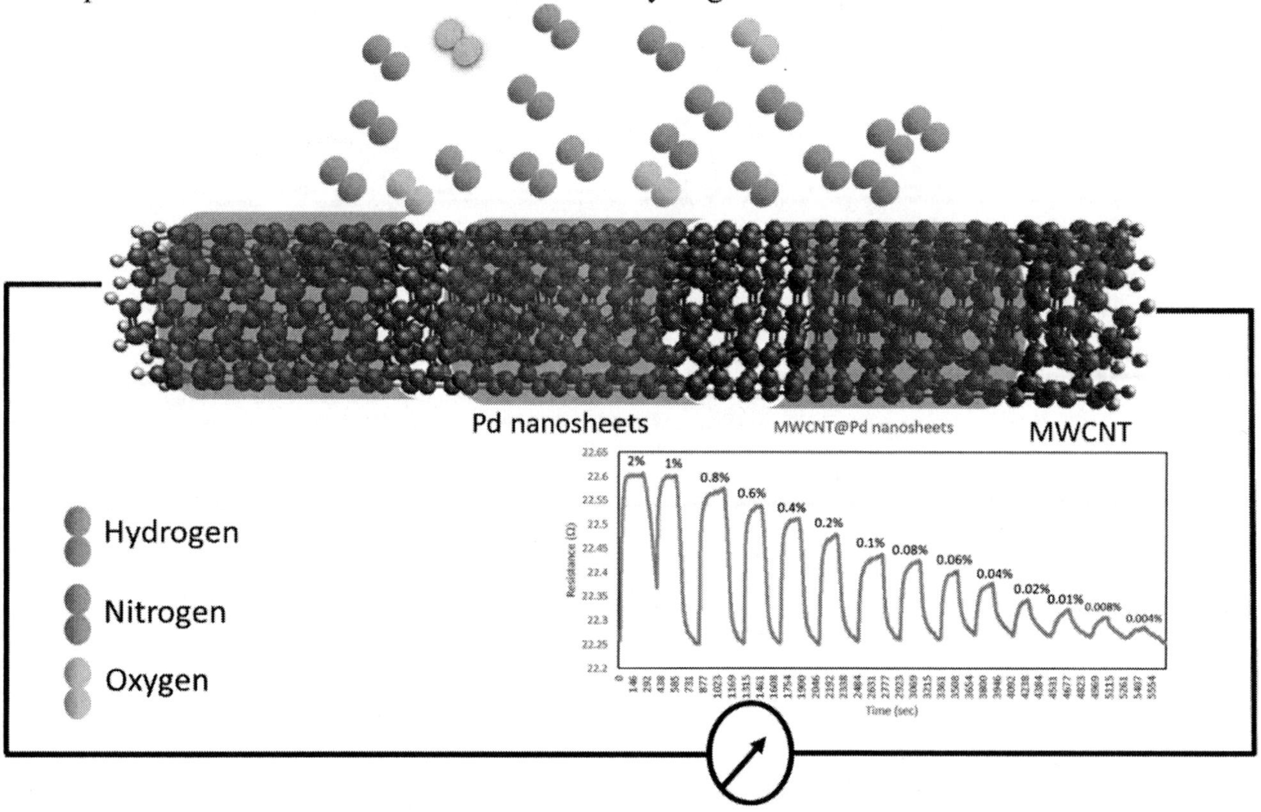

608092 Flame Aerosol Synthesis of Palladium-Decorated Crumpled Reduced Graphene Oxide Nanocomposites for Hydrogen Detection at Room Temperature

Abstract:

Hydrogen (H_2), a green energy source, is highly flammable with a lower explosive limit of just 4% in air. Its leakage during production, transportation, and storage can result in disastrous consequences. Therefore, development of accurate, rapid, and stable H_2 sensors that work in air at room temperature is crucial. In this presentation, we will describe a novel three-dimensional palladium (Pd)-decorated crumpled reduced graphene oxide ball (Pd-CGB) nanocomposite for H_2 detection in air at room temperature. Pd-CGB nanocomposites were synthesized using a flame-driven high temperature reducing jet (HTRJ) process. In this process, a hydrogen-rich flame passes through a converging-diverging nozzle. An aqueous precursor solution containing graphene oxide and palladium nitrate is injected at the throat section of the nozzle and is atomized by the high velocity gas stream. The resulting droplets evaporate, graphene oxide sheets crumple, and the palladium precursor decomposes, initiating nucleation of palladium nanoparticles on crumpled graphene oxide in a reducing environment containing excess H_2. After the reaction zone, nanocomposites are cooled immediately to prevent further growth and coalescence of the Pd nanoparticles. Hydrogen sensors produced by simple drop-casting of these nanocomposites onto electrodes performed well over a wide range of H_2 concentrations (0.01-2%) with response (resistance change) value, response time, and recovery time of 14.8%, 73 s, and 126 s, respectively, at 2% H_2. Compared with other reports on Pd-graphene sensors, the present study shows superior response/recovery time. The HTRJ process enables the single-step, continuous production of Pd- and other metal-decorated graphene nanocomposites with great potential for creating various gas sensors by simple drop-casting onto low-cost electrodes.

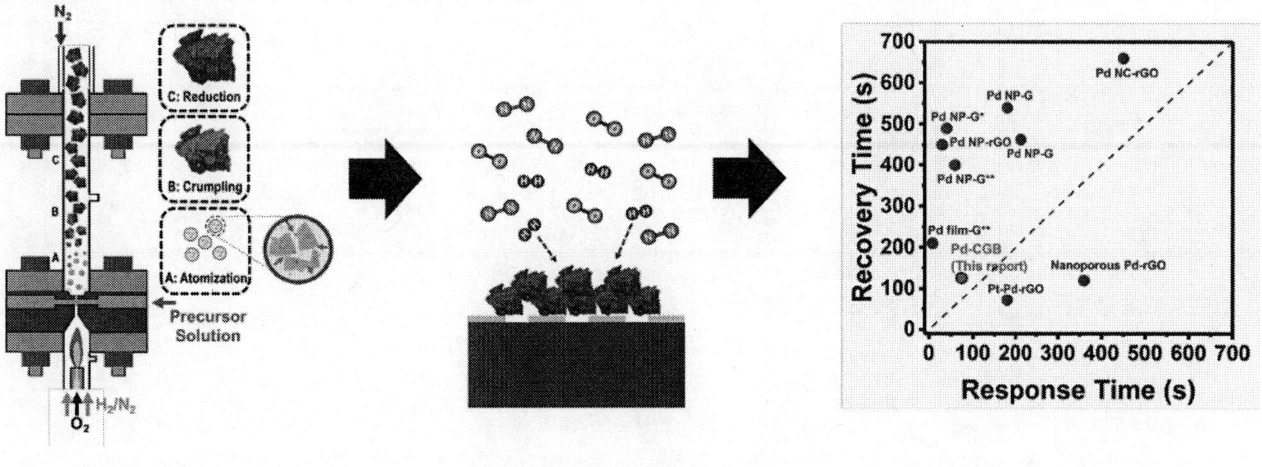

601668 Carbon Black-Gold Nanoparticles for Detection of Analytes Using Surface Enhanced Raman Scattering

Abstract:

Surface-enhanced Raman scattering (SERS) is a powerful technique for the detection of a variety of molecules at very low concentrations. We have developed irregularly shaped gold nanoparticles using a carboxyl-terminated carbon black (CB) nanoparticles as a template. The CB particles were coated with the cationic polyelectrolyte poly-L-lysine (PLL) and exposed to a tetrachloroauric acid solution. Gold - carbon black (AuCB) particles were formed by the reduction of gold chloride ions that concentrated on the surfaces of the PLL. The morphology of the gold on the AuCB can be modified using different PLL concentrations,. The plasmon resonance can then be matched to the incident laser wavelength, thus greatly enhancing Raman signals coming from analytes in the vicinity of the gold surface. We used these AuCB particles to detect a wide range of analytes, including 4-nitrobenzene thiol, and nitrate and sulfate ions in a mixture. The limits of detection were well below that obtained by other SERS particles.

610202 Molecular Recognition and *In Vivo* Detection of Temozolomide and 5-Aminoimidazole-4-Carboxamide for Glioblastoma Using Fluorescent Nanosensors

Abstract:

The efficacy of chemotherapeutics such as temozolomide (TMZ) and its metabolic product 5-aminoimidazole-4-carboxamide (AIC) is often affected by the timing, quantity and frequency of dosages. There is strong interest in facilitating the ability to monitor efficacy in individual patients for specific subtypes of cancer. Real-time, dynamic measures of potency may also supplement or in some cases replace reliance on bio-imaging. Towards this end, in this work we develop new, synthetic molecular recognition sites for TMZ, and its decomposition product AIC, grafting them onto near infrared fluorescent nanoparticles capable of forming optode or other biosensor interfaces to monitor drug efficacy in real-time. Infrared fluorescent single-walled carbon nanotubes, wrapped using specific DNA oligonucleotides, are encapsulated within biocompatible, poly(ethylene glycol) diacrylate hydrogels and enable the selective recognition of an anti-cancer drug, TMZ, on U-87 MG human glioblastoma cells. In both solution phase and hydrogel form, the sensors exhibit a fluorescence response to TMZ with a detection limit of 30 M. Furthermore, U-87 cells exhibit no changes in viability for 7 days in contact with the hydrogel. The sensors were used to track the progression of glioblastoma death following TMZ administration. In addition, the sensors were able to monitor *in vivo* AIC and TMZ dynamics in SKH-1E mice. These results enable real-time detection of chemotherapeutic concentration and metabolism at the cellular and subcellular levels, with potential to increase the efficacy of cancer treatments.

606459 An Enzymatic Electrochemical Biosensor for Real-Time Detection of Physiologically Relevant Nicotine Concentrations

Abstract:

The glucose biosensor, built upon the redox enzyme glucose oxidase, is the most commercially successful and studied enzymatic biosensor. However, the lack of available and functionally validated enzymes is prohibiting the development of redox-based sensors for other important analytes. Herein, we present the development and assessment of an electrochemical nicotine biosensor, using genomic screening to identify the gene for a known nicotine catabolizing redox enzyme. The resulting nicotine biosensor demonstrated a specific, sensitive, and stable operational profile with a limit of detection of 27 μM over the range of 0-200 μM. This range is well within the physiological concentrations of nicotine present in smoker urine. Specificity and cross-reactivity were measured against structurally similar compounds to nicotine as well as to known physiological by-products. Our results highlight that this novel enzymatic electrochemical nicotine biosensor possesses operational capabilities for monitoring of nicotine in physiologically relevant conditions. The screening methodology can be generalized for the discovery of enzymes for novel sensor development.

608267 Simulations Guide Optimization of Electroenzymatic Biosensors for Neurotransmitters and Enable Proper Interpretation of Sensor Response *In Vivo*

Abstract:

Electroenzymatic biosensors for neurotransmitters are uniquely compact, sensitive, and selective devices that provide the neuroscience community with a method to monitor fluctuating chemical concentrations in the brain. Many studies primarily consider glutamate, the most common excitatory neurotransmitter, and study the relation between sub-second neurotransmitter dynamics and various stimuli, behaviors, or diseases. Challenges, however, have remained in the optimization of sensor design and in clarifying the significance of assumptions made during *in vivo* use. Accordingly, a number of simulations were carried out to guide the development of glutamate sensors and to test the limitations of the most prominent assumption, that sensor response can be linked to the immediate extracellular neurotransmitter concentration using a predetermined calibration factor. Recommended modifications to sensor design were developed and successfully implemented, improving sensitivity six-fold (up to 320 nA/μM/cm^2) with an accompanying order of magnitude decrease in response time (down to ~80 ms). Since even this improved response time is an order of magnitude slower than the timescale of synaptic neurotransmitter release, it is clearly wrong in many cases to assume that the use of a calibration factor to translate sensor response to neurotransmitter concentration without any signal deconvolution is accurate. To investigate the significance of this assumption and the conditions for which this practice remains reliable, the model was expanded with a new set of partial differential equations to describe the extracellular space of the brain. Considering the effects of variations in the local biological environment bordering a sensor, a range of representative parameters for some biological processes were selected. Comparisons were made between the theoretical responses expected from *in vitro* calibrations, tonic steady-state glutamate presence, and bolus-type glutamate releases that mimic synaptic release. Surprisingly, in steady-state cases, the sensor response could indicate a concentration over twice the actual value, while in cases where sensors were 10 μm or more from the source of glutamate being detected, sensor response was < 50% of the expected value. In some cases, transient concentrations due to a single synaptic-type release of neurotransmitters at a moderate distance from the sensor were not detected by the sensor at all, and when they were, the time course of signal decay did not match the rates of reuptake expected in the brain without significant signal deconvolution. Ultimately, results show that additional characterization of sensor performance *in vivo* may be necessary to enable proper interpretation of sensor data.

611068 Theoretical and Experimental Studies on an Electrochemical Enzyme Immunosorbent Biosensor

Abstract:

Electrochemical immunosensors integrate biological recognition molecules (e.g., antibodies) with redox enzymes (e.g., horseradish peroxidase (HRP) to combine the advantages of immunoassays (extremely high sensitivity and selectivity) with those of electrochemical biosensors (reproducible, quantitative electrical output). Mechanistic mathematical models that describe the multiple mass-transfer and chemical reaction steps that give rise to the electrical output are needed to help design, optimize, and validate electrochemical immunosensors for medical applications. In this study, we developed such a model of an electrochemical immunosensor that used HRP as a reporter enzyme on the secondary and validated it experimentally.

The model consists of a system of non-linear mass-balance equations that describe simultaneous mass transfer and chemical reactions of the reactants (catechol and hydrogen peroxide) that HRP oxidizes, as well as the HRP product, O-quinone, which is reduced at the working electrode to give the amperometric output. A bi-substrate, ping-pong reaction mechanism was used to describe the enzymatic reaction, and the *Butler-Volmer* kinetic model was used to describe the reduction of o-quinone back to catechol at the working electrode's surface. The system of equations was solved numerically using the BVP4C function in MATLAB.

Experimental data for model validation was obtained using electrochemical immunosensors for a surrogate protein antigen (mouse IgG). The immunogens interface was fabricated on the gold working electrode of DropSens screen printed electrode arrays. The steady-state amperometric output of the resulting immunosensors was measured over a range of four independent variables: the concentrations of the two reactants, the voltage of the working electrode, and pH. The experimental results were used to determine optimal operating conditions and to validate the model. The validated model was then used to calculate Damkohler numbers and flux-control coefficients to identify mass-transfer and reaction steps that most strongly affected the magnitude of the immunosensors' output and the output's sensitivity to analyte concentration.

610485 HTL Derived Biochar and Graphene Nanoplatelets for Biosensor Applications

Abstract:

HTL Derived Biochar and Graphene Nanoplatelets for Biosensor Application

Bharath K Maddipudi, Vinod Amar, Hope Dosch, Anuradha Shende and Rajesh V Shende*Department of Chemical and Biological Engineering, South Dakota School of Mines and Technology, Rapid City, South Dakota 57701, United States of America

***Corresponding Author:** Rajesh Shende, Department of Chemical and Biological Engineering, South Dakota School of Mines and Technology, Rapid City, South Dakota 57701, United States of America (E-mail: *Rajesh.Shende@sdsmt.edu*)

Abstract

Biosensors with Au electrode layer immobilized with peptides are currently being developed for the detection of microbial contamination. It is anticipated that the use of a low-*k* dielectric layer modified with graphene will improve the sensor sensitivity. The low-*k* dielectric layer can be modified with the biochar derived from hydrothermal liquefaction of different biomass substrates. In this investigation, HTL of lignocellulosic substrates was performed under non-catalytic and catalytic conditions and by-products were analyzed. The biochar was thermally treated in a tubular reactor under inert environment from 400-1100°C without any pretreatment. In other set of experiments, HTL derived biochar was acid treated and subsequently, thermal activation was performed. After the acid pretreatment and thermal treatment steps, the porosity, the specific surface area (SSA) and the morphology of the char was analyzed using BET and SEM/EDX. The char obtained after thermal treatment was combined with low-*k* dielectric layer and biosensor was fabricated. The sensor output was compared with the biosensor fabricated with low-*k* and graphene nanoplatelets. In addition, the sensor output was correlated with the char materials obtained under different thermal conditions. The results obtained on hydrothermal liquefaction of biomass substrates, characteristics of biochar before and after thermal treatment, biosensor fabrication and sensor output will be presented.

604780 Construction of a Red Emission Bodipy-Based Probe for Tracing Lysosomalviscosity Changes in Culture Cells

Abstract:

A red emission BODIPY-based lysosome-targeted viscosity probe (Lys-VBOD) was synthesized for monitoring lysosomal viscosity in live cells. The increase of viscosity can restrict the twisting of double bond between indole
and BODIPY fluorophore, also restrict the rotating of single bond between quinoline and BODIPY, which lead to the enhancement of fluorescence intensity in high viscosity environment. The changes of fluorescence intensity
exhibit a good linear relationship with viscosity values. Besides, Lys-VBOD is quite stable in various pH solvents with given viscosity, suggesting that Lys-VBOD is suitable for applying in a wide pH range. Moreover, Lys-VBOD
can visualize the changes of lysosomal viscosity in live cells under the stimulus of dexamethasone. BSA titration experiment demonstrated that Lys-VBOD is stable in the presence of macromolecules and proteins. Notably, the
fluorescence lifetime of Lys-VBOD in large viscosity solvents is extremely high. These results indicate that LysVBOD is an excellent tool for detecing viscosity changes in vitro and live cells.

610593 Super Sensitive Cerium Oxide-Based Composite Sensor for the Detection of Hydroxyl Radicals

Abstract:

Free radicals are recognized as essential molecules for upholding the function of normal cells. Immune systems, such as bacteria elimination and virus inactivation, are examples of the crucial role that free radicals play in human cells. Even though free radicals have benefits, an optimum production is key to maintain their advantages, as an excess can cause the development of oxidative stress conditions, such as cancer, Alzheimer's, and Parkinson's diseases. Among all free radical species, hydroxyl radicals (•OH) are the most reactive and dangerous, and they can be used as biomarkers for the detection of those diseases at their initial stages. Thus, a very sensitive analytical sensor is needed to detect and monitor low concentrations of •OH. The integration of an electrochemical technique with a sensing element is regarded as a promising method for •OH detection due to a rapid and direct measurement without the pretreatment of samples. Thus, in this study, a glassy carbon electrode was modified with a sensing element comprised of ultrasmall cerium oxide nanoclusters (<2 nm) grafted onto a superconductive carbon (Figure 1a). XPS results showed 46.3% Ce^{3+} sites, which are those able to react with •OH. Cyclic voltammetry (CV) was implemented to characterize the interaction of the composite sensor with •OH, and they showed that the composite sensor was able to differentiate between solutions with and without •OH by an increased redox response in presence of •OH (Figure 1b). Furthermore, we observed a linear relationship between the redox response and the concentration of •OH in the range of 0.1 mM to 10 mM, with a limit of detection as low as 0.1 μM when using only 0.5 wt% CeO_x loading.[1]

(1) A.C. Alba-Rubio, D.S. Kim, S. Duanghathaipornsuk; Super sensitive sensor for the detection of hydroxyl free radicals with scavenging properties. Application number 62/977,945 (2020).

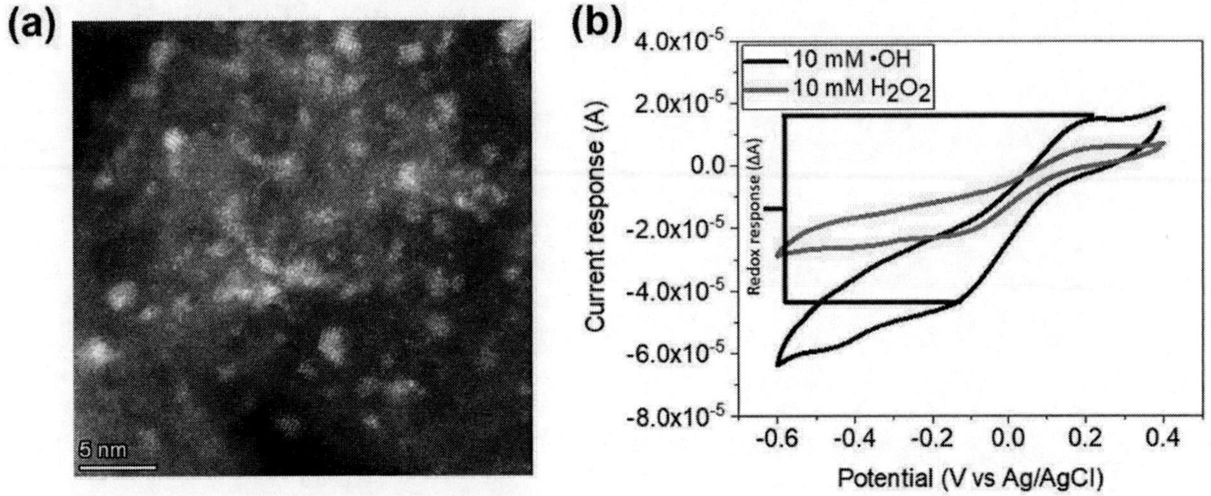

Figure 1. (a) STEM image of 0.5 wt% CeO_x/Carbon and (b) Current response to hydroxyl radicals.

611380 High Throughput Antibiotic Susceptibility Testing with Optical Nanosensors

Abstract:

Infections in a clinical setting can result from biofilm-forming species, which presents two key challenges: increased antimicrobial resistance, and poor diagnostic approaches to categorize antimicrobial susceptibility. Traditional antimicrobial susceptibility assay approaches are built upon planktonic bacteria (bacteria that are growing in solution). Biofilms are bacteria that grow in a community on a surface, and this community structure changes resistance to antimicrobials that is not reflected in these planktonic assays. While current biofilm-based assays exist, they are both time and material intensive, and can still provide antibiotic susceptibility results that may not reflect the actual biofilm response. Here we present our nanosensor based approach to monitor biofilm metabolic response during the administration of antibiotics. We grow biofilms with oxygen-monitoring nanosensors embedded throughout the film, and these nanosensors continuously report out the metabolic state of the biofilm. As we add antibiotics, we can measure what concentration can stop metabolism, and infer the minimum biofilm inhibitory concentration (MBIC) of the specific strain. Our diagnostic approach functions in 96 well plates with minimal materials and time (important for high throughput screening), and works with both lab and clinically relevant strains (demonstrated with Pseudomonas aeruginosa).

597841 Surpassing the Detection Limit and Accuracy of Electrochemical DNA Sensor through the Application of CRISPR-Cas Systems

Abstract:

We present a CRISPR Cas system enhanced electrochemical DNA (E-DNA) sensor with unprecedented sensitivity and accuracy. The principle of the E-DNA sensor is the target induced conformational change of the surface signaling probe (containing an electrochemical tag), leading to the variation of the electron transfer rate of the electrochemical tag. With the introduction of CRISPR cleavage activity into the E-DNA sensor, a more apparent signal change between with and without the presence of the target can be achieved. We compared the performance of Cas9 and Cas12a enhanced E-DNA sensor and optimized the chemical environment of CRISPR, achieving a femto-molar detection limit without enzymatic amplification. Moreover, we correlated the CRISPR cleavage signal with the original E-DNA signal as a strategy to indicate potential mismatches in the target sequence. Comparing with classic DNA electrochemistry based mutation detection strategy, CRISPR enhanced E-DNA sensor can determine the presence of a single mutation at an unknown concentration condition. Overall, we believe that the CRIPSR enhanced E-DNA sensing strategy will be of especially high utility for point-of-care systems owing to the programmability, high-sensitivity and high-accuracy.

592702 In Planta Detection of Synthetic Auxin Plant Hormones Using Single-Walled Carbon Nanotubes (SWNT) Based Nano-Sensors

Abstract:

1-Naphthalene acetic acid (NAA) is one of the most commonly-used synthetic auxin plant hormones in agriculture as rooting agent or as sprays to prevent premature flowering and fruit drop. NAA is subject in some countries to regulatory control, requiring monitoring of NAA levels in vegetables and fruits. Unfortunately, state-of-the-art NAA sensing methods are destructive and laborious. In this work, we show the development of a novel NAA nano-sensor based on the concept of corona phase molecular recognition, using cationic polymer wrapped single-walled carbon nanotubes (SWNTs). This approach allows real-time non-destructive *in planta* detection of NAA. Using a semi-rational approach to polymer design, we developed a sensor that is selective for auxins, with particularly large responses to NAA as well as indole-3-butyric acid (IBA), both of which are used as rooting agents. An unresponsive control sensor was also developed to enable a ratiometric sensing platform. The sensor is demonstrated *in planta* for living spinach plants. We further probed the corona phases of both the active and control sensors, using the molecular probe adsorption (MPA) method to measure their accessible surface area for analyte adsorption. In both sensors, the Gibbs free energy of the riboflavin molecular probe binding to SWNT scales inversely with cube root of accessible surface area, which is in agreement with the structure-property relationship established for corona structures across different systems.

Since its initial discovery and commercialization in the 1940s, 2,4-Dichlorophenoxyacetic acid (2,4-D) has been an selective auxin herbicide commonly used for agricultural purposes. With increased usage of 2,4-D, herbicide resistance in weeds is of concern. To date, there are 28 reported species of weeds resistant to 2,4-D. Application of ineffective herbicides with poor levels of control causes wastage of money to farmers and detriment to the environment. In this work, we have developed a selective SWNT-based nano-sensor for 2,4-D based on the same corona phase molecular recognition approach, which also allows real-time non-destructive *in planta* detection of 2,4-D. This helps prevent wasteful spraying and ineffective deployment of herbicides onto resistant weeds. It also allows us to calibrate the herbicide dosage to minimize its usage. Further, it provides a rapid screen for herbicide susceptibility without need to monitor plant growth over days. The sensor is demonstrated *in planta* for living non-model plant species such as spinach, pak choy (*Brassica rapa subsp. chinensis*) and choy sum (*Brassica chinensis var. parachinensis*).

610521 *Electrochemical Sensors for Microbial Activities in Benthic Sediments: A Sentry for Lacustrine P Biogeochemistry*

Abstract:

Internal loading (i.e. release from benthic sediments) of PO_4^{3-} to the lacustrine water column could induce harmful algal blooms (HABs). The extents of internal PO_4^{3-} loading are difficult to constrain and detect in situ and in real time because fluxes are measured in laboratory incubations, monitoring approaches are difficult to deploy in the field, and the scales of microbially-induced redox gradients can occur on sub-centimeter scales. Release or entrained of PO_4^{3-} from benthic sediments is controlled by the activities of microorganisms in those sediments. In this work, we develop an electrochemical approach to detect, identify, and monitor benthic sediment-associated microbiological activities that could be deployed in situ and in real-time.

A series of laboratory and field experiments was conducted to identify electrochemical signatures associated with a variety of microbiological activities occurring in Lake Erie sediments. Our technique makes use of zero-resistance ammetry (ZRA) measurements, which we have previously used to diagnose microbiological activities. The approach uses inexpensive materials, demands little power, and can interrogate microbial activities at small spatial scales. The ZRA approach detects internal PO_4^{3-} loading (the release of PO_4^{3-}, as "dissolved reactive phosphorous" from sediments), which can contribute to harmful algal blooms (HABs). Microbial degradation of algal biomass and microbially-induced mineralogical transformations can profoundly influence the PO_4^{3-} concentration in the water column, and therefore, the potential for HABs despite adequate management of external PO_4^{3-} loading. Thus, microbial metabolism in PO_4^{3-}-loaded benthic sediments represents a remnant threat to induce or contribute to HABs well into the future. Microbiological activities induce redox gradients in sediments, depending on the predominant terminal electron accepting process. Most notable from the standpoint of PO_4^{3-} flux is the anaerobic reductive solubilization of Fe(III) (hydr)oxide phases by Fe(III) reducing bacteria (FeRB), which releases PO_4^{3-} that was previously adsorbed to those phases. However, microbial activity can induce a variety of other chemical conditions that enhance or limit PO_4^{3-} solubility (e.g. Fe(II) oxidation, organic carbon mineralization, acid production, primary productivity). *ZRA-based monitoring can determine extents and types of microbial metabolism in the sediments, be exploited for long-term stewardship of Lake Erie sediments, and serve as an "early-warning system" for internal PO_4^{3-} loading events.*

This presentation describes our newly-developed, ZRA-based technique for monitoring sediment activity and internal PO_4^{3-} loads. We believe that this will eventually be an "early warning system" for internal PO_4^{3-}-supported HABs.

608727 Iot Based Fuel Adulteration Detection Using Dispersion Model and Tail Pipe Emissions

Abstract:

Over the years up-gradation to cleaner fuel aims to reduce the rising environmental pollution. Evidence from the past states that adulteration of fuel exacerbates the problem. To cut the cost of highly taxed improved hydrocarbon, it is usually blended up to 80% with cheaper hydrocarbon. The detection of adulteration is to be achieved by implanting sensors within the fuel tank and the exhaust tailpipe. This whole process will be brained by the Engine Control Unit (ECU) and the data will be transmitted to the pollution monitoring agencies via Internet of Things (IoT) over an interval of time. The data received from the vehicle will be modeled and analyzed on the basis of the dispersion model. The molecular diffusion of the adulterate in the quality hydrocarbon will result in the intermixing of molecules, by extending this study to reverse engineering, we will determine the amount of adulterate injected. The data will be assimilated with the fuel and emission standards that will estimate the quantity and the injection time period of the adulterate. Looking closer to the data, it can be easily estimated the adulteration cases over a zone by analyzing the chronicle of refueling. This ensures the environmental agencies to keep track of the adulterate hotspots. Presently various testing parameter are available such as ASTM D6730 (Gas Chromatography), ASTM D2425(Mass Spectroscopy), ASTM D2425 (Distillation). The common limitation of the tests available is that analysis must be performed outside the vehicle body, that will no longer provide the chronology behind the addition of adulterant. It is estimated that when this model will be implemented to the vehicles, it will uplift the aspect of the cleaner fuel to deliver better engine performance with reduced emission levels. This will empower every citizen to monitor the quality of fuel easily by accessing the data from the vehicle.

605985 In-Situ Chemical Composition Sensing Using IR Adsorption in Micro-Structured Flow Reactors

Abstract:

Flow reactors are attractive options for moving beyond small-scale batch synthesis for some high-value chemicals, particularly in the fine chemical industry. One major limitation at the present is the lack of online sensors, for example it is challenging to design a reactor to carry out a complex organic chemical reaction, since most available analytics require sampling and relatively slow, offline analysis, making the results of limited use for process control.

In our DFG-funded Research Unit called Promise (https://promise.kit.edu) we explore multiple options for the in-situ sensing and control of a range of different physical and chemical processes. One project in particular explores the possibility for using relatively low-cost, solid-state IR sensors embedded into a flow reactor along the flow path, to give real-time, spatially resolved measurements of chemical compositions, which can be correlated to the extent of the chemical reaction (and even more complex measures like product selectivity, by-products and so on). In this presentation we outline our latest reactor design, which features three distinct and individually temperature-controlled reaction zones, two IR sensing zones (with specific materials of construction that allow for IR-transmission over a very wide range of wavelengths, much greater than can be achieved using conventional glass materials), and unique flow paths allowing for full system sealing at the top and bottom of the reactor. This apparatus was constructed in the micro process engineering workshop and combined with unique and custom-built IR sensors, resulting in a system that can make real-time IR analysis and use the results for process monitoring and eventually process control.

The Figure shows exploded views (from the top and bottom), indicating the complexity of the device and the careful design decisions necessary to achieve optical transparency in all required regions, while meeting flow and sealing requirements. Initial results for several photochemical reactions will be presented.

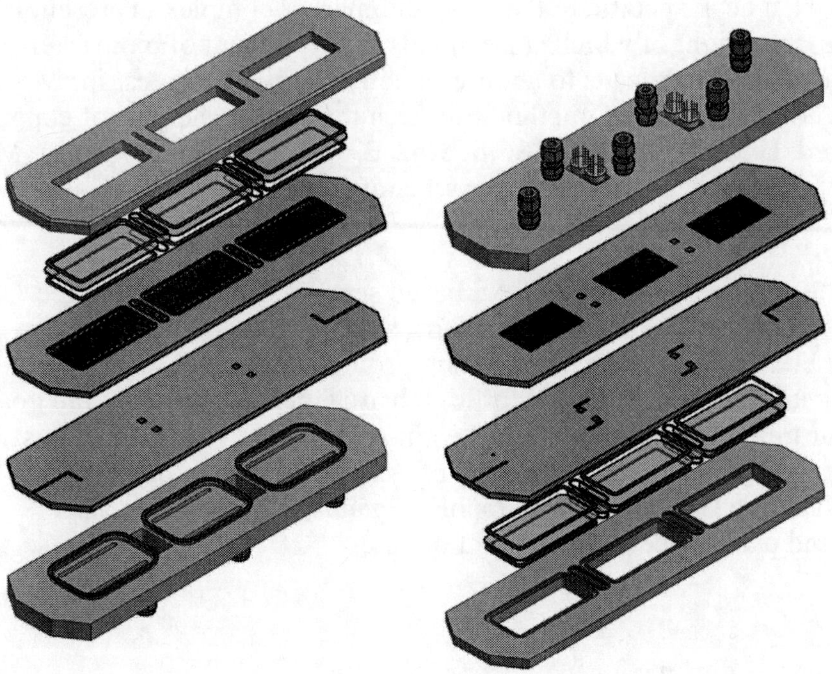

611466 Rapid Detection of Tuberculosis Breath Biomarkers Using Engineered Electroactive Solutions (EAS)

Abstract:

The World Health Organization has designated the development of a rapid, inexpensive, point-of-care screening test for Tuberculosis (TB) as high priority in their annual Global Tuberculosis report, reporting nearly 1.5 million TB related deaths in 2019 [1]. Screening via associated breath biomarkers has been identified as a promising emerging technology faster and easier to use at the point of care than standard sputum-based tests. Traditional GC-MS testing has identified methyl nicotinate (MN) and methyl p-anisate (MPA) as specific biomarkers of mycobacterium tuberculosis that are not readily present in ambient air or the breath of healthy patients, however this mode of testing is prone to confounding and is often prohibitive at the point of care due to its high cost and need for highly skilled technicians. The relatively high solubility of these semi-volatile compounds in aqueous solution provides a unique ability to isolate the biomarkers from more volatile breath components by a diffusive gas to liquid transfer process. This step serves to both purify and effectively concentrate the sample for electrochemical analysis.

Previously, our lab developed a sensing platform utilizing Engineered Electroactive Solutions (EAS) on inexpensive, commercially available screen-printed electrodes, in which the change in a transition metal of known electrochemical activity is observed when exposed to an analyte of interest [2]. The original study was successful in using a divalent copper electroactive solution (Cu2-EAS) to detect MN down to 1 mM using cyclic and square wave voltammetry techniques (CV and SWV, respectively). However, the relatively high reactivity of copper may limit specificity. The sensing platform has been improved to use a divalent cobalt electroactive solution (Co2-EAS), with a limit of detection of about 0.01 mM of MN, about 100x more sensitive than Cu2-EAS. When Co2-EAS is mixed with a sample solution of MN, a peak specific to the metal-biomarker complex is present in both the forward and reverse scans, the height of which corresponds to the amount of biomarker present (see figure 1). The addition of this peak leads to a more sensitive and specific test for MN than the Cu2-EAS, The Cu2-EAS utilized a “fingerprint” method in which the relative heights of copper’s existing redox peaks were compared. Unlike the Cu2-EAS, the Co2-EAS is also able to identify MPA in solution by a significant change in the existing cobalt peaks when it is added to solution, as shown in figure 2.

The simple sample collection and testing methods for an EAS based sensor is ideal for resource limited communities where TB is prevalent. The ability to perform sample collection and testing remotely as well as store collected breath samples for later testing provides flexibility to both patients and technicians. Disposable testing kits can be assembled for less than $5 per test, meeting the goal price point for TB triage testing put forth by Van’t Hoog et al, estimating a 42% reduction in cost compared to “Xpert for all” testing in Uganda [3]. Field testing of this novel sensing platform will be conducted in collaboration with Makerere University in Kampala, Uganda to determine sensor performance and to better understand patient needs at the point of care.

References:

[1] World Health Organization, “Status of the TB epidemic,” 2019.

[2] Willis C, Misra M, Mohanty S, 2019, “Engineered Electroactive Solutions for Electrochemical

Detection of Volatile Organic Biomarkers Associated with Disease," 235th *ECS Meeting*, Dallas TX, 26-30 May

[3] A. H. van't Hoog et al., "Optimal Triage Test Characteristics to Improve the Cost-Effectiveness of the Xpert MTB/RIF Assay for TB Diagnosis: A Decision Analysis," PLoS One, vol. 8, no. 12, p. e82786, Dec. 2013.

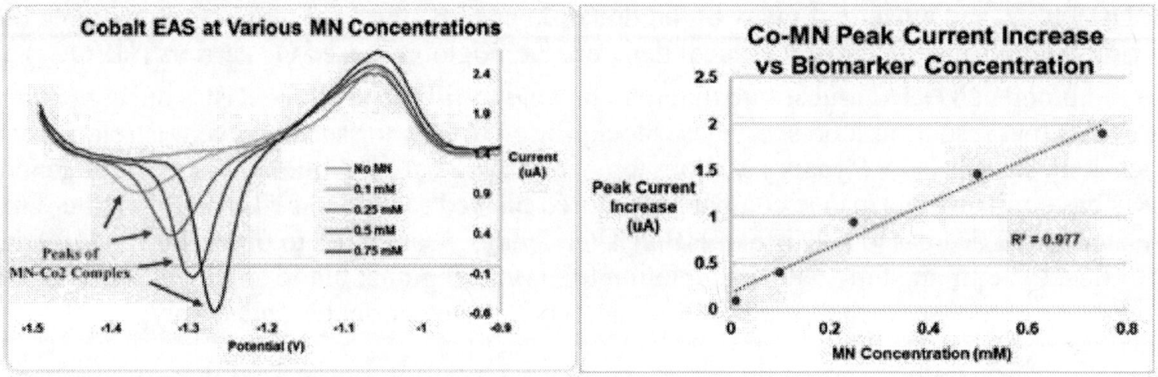

Figure 1. Size of peak corresponding to MN-Co2 complex is proportional to concentration of MN in solution

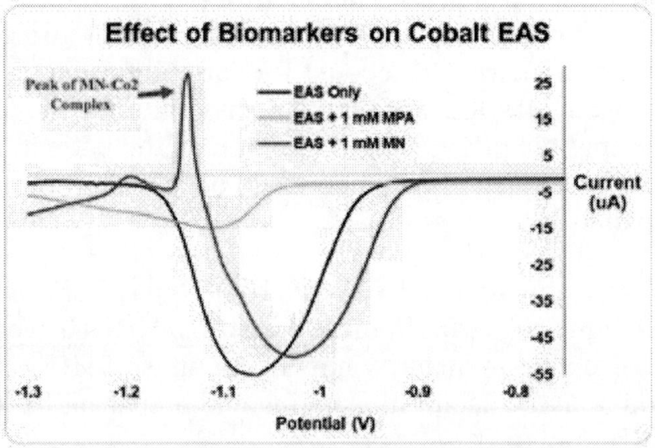

Figure 2. Comparison of square wave voltammograms of Co2-EAS with and without biomarkers present

607149 Exploring the Potential Role of Hemoglobin-Based O_2 Carriers in a Pandemic Simulation of the Blood Supply Chain

Abstract:

Introduction: The recent outbreak of coronavirus disease 2019 (COVID-19) has a significant effect on regional and national blood supply chains. Due to decreased donor availability, inadequate access to COVID-19 tests, and closure of many blood donation centers, the total number of available blood units in circulation is unable to meet clinical demand. Hemoglobin-based O_2 carriers (HBOCs) are a potential red blood cell (RBC) substitute that may be able to fill the need for RBCs during pandemics or other events that result in blood shortages. Stockpiling HBOCs for such an event is relatively easy given their long shelf life (> 6 years) at room temperature (25 °C). Although the current generation of HBOCs has similar performance compared to stored packed red blood cells (pRBCs), there may be issues related to the decreased half-life of HBOCs (1-2 days) compared to pRBCs (50 – 100 days). In light of this, the current study uses a simulation of an interregional blood supply chain in a pandemic scenario to assess what role HBOCs may have in the national blood supply.

Methods: The current study uses an interregional stock-and flow simulation of the national blood supply chain in the United States. The simulation tracks the quantity and blood type in blood banks and hospitals in 4 regions of the United States as defined by America's Blood Centers (ABC). Blood units are collected and transferred from blood banks to hospitals within their respective region. A pandemic simulation was performed by reducing the rate of blood donation within each region. Varying reduction and time for recovery were used as metrics to account for variability in the incubation period and duration of the given pathogen. HBOCs were implemented into the model as an RBC substitute that would only be used if no compatible pRBC unit is available to the patient. A half-life based algorithm was used to determine what percentage of patients would need an additional supplement of HBOC due to its relatively low half-life.

Results and Conclusion: With this model, we were able to quantify how HBOCs may help reduce strain on the blood banking system during a pandemic scenario. In these scenarios, HBOCs were able to attenuate the rate of pRBC deficits. We were also able to quantify how many units of HBOCs may be needed in future pandemic scenarios. This tool may be useful in planning and preparing for future pandemic scenarios after HBOCs obtain clinical approval.

611244 Copper (II) Oxide Nanomaterial for Electroactive Inactivation of Coronavirus in Aerosols and Liquids

Abstract:

Viruses, more specifically coronaviruses, are extremely contagious and are transmitted through physical contact and airborne microdroplets. The recent SARS-CoV-2 (COVID19) has high surface stability and can remain suspended in the air for several minutes. These characteristics make disinfecting the environment difficult. Common viral disinfection methods are chemical disinfectants (chlorine and alcohol enriched products), ozonation, and ultraviolet light. These disinfecting methods can be expensive or form toxic byproducts. In this work, we present a novel approach to deactivating viruses, employing a nanostructured copper(II) oxide platform under given potentials. Prior studies have shown the inactivation of viruses on copper and copper ions surfaces[1-3]. A recent study also showed that the decay of COVID19 on copper surfaces was faster than other surfaces such as steel, plastic, and cardboard[4].

Copper(II) ions, being highly reactive, are a good candidate for inactivating viruses. The nanostructured copper(II) oxide can be integrated into various devices suitable for cleaning the environment such as a surface cleaning wand or air purifier. It can also be integrated into personal protective equipment, such as masks worn by first responders. In this work, copper foil was anodized to create copper oxide nanostructures. Various annealing conditions in oxygen were explored to create a surface enriched with copper(II), which we hypothesize will be the most effective in the inactivation of viruses[1]. Coronaviruses were introduced to the conductive copper oxide nanostructure under various potentials to induce electrochemical reactions (such as oxidizing the virus RNA). Betacoronavirus, Avian Coronavirus, and Transmissible Gastroenteritis virus were chosen as surrogates for SARS-CoV-2, tested on the cupric oxide nanostructure. The final goal of this work is to utilize the nanostructure material in an electroactive device, designed to inactive coronavirus immediately upon contact. The applied potential to the copper oxide nanostructure results in instantaneously inactivation of the virus, compared to commonly viral inactivation methods that needed extended contact time. Results on the inactivation of the mentioned viruses will be presented.

Figure (1) is an SEM picture of copper oxide nanostructures made in a mixture of , NaOH and ethylene glycol. Data on anodized copper foil in solutions containing various amounts of the mentioned mixtures such as additive of fluoride ions to get to a sufficient amount of copper(II) oxide will be presented. The oxidative state of the copper oxide nanomaterial will also be explained during the presentation by showing X-ray photoelectron spectroscopy analysis.

References:

[1] M. Horie *et al.*, "Inactivation and morphological changes of avian influenza virus by copper ions," *Arch. Virol.*, vol. 153, no. 8, pp. 1467–1472, 2008.

[2] Sundberg, K.; Champagne, V.; McNally, B.; Helfritch, D.; Sisson, R., Effectiveness of nanomaterial copper cold spray surfaces on inactivation of influenza A virus. J Biotechnol Biomater 2015, 5 (205), 2.

[3] Longano, D.; Ditaranto, N.; Sabbatini, L.; Torsi, L.; Cioffi, N., Synthesis and antimicrobial activity of copper nanomaterials. In Nano-Antimicrobials, Springer: 2012; pp 85-117.

[4] D. Taylor, A. C. Lindsay, and J. P. Halcox, "c or r e sp ondence Aerosol and Surface Stability of SARS-CoV-2 as Compared with SARS-CoV-1," pp. 0–2, 2010.

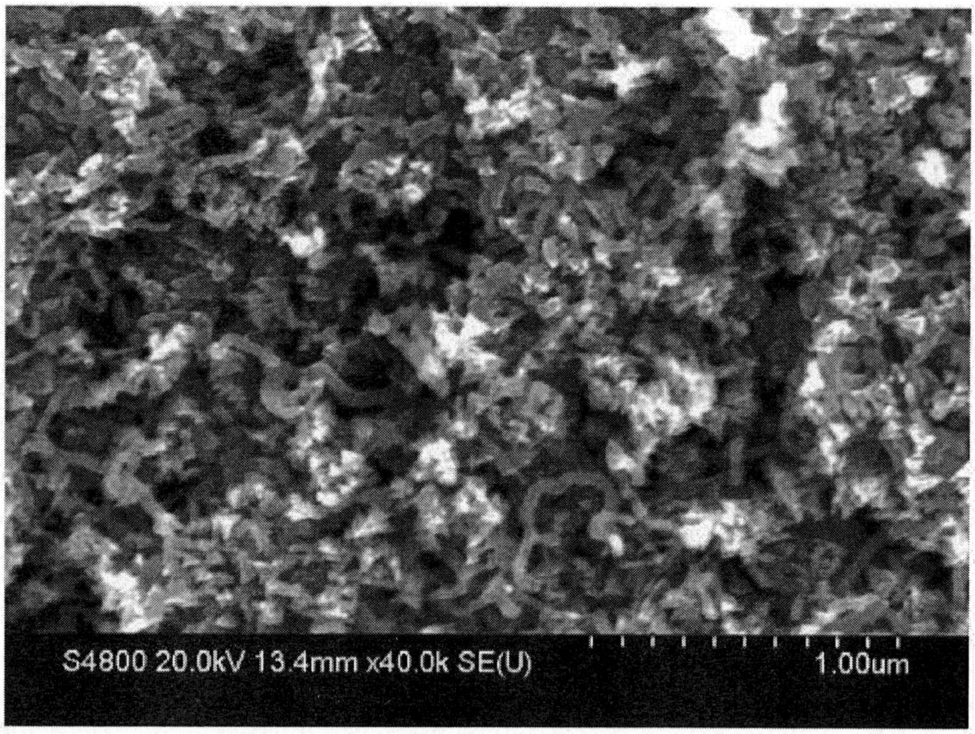

Figure (1), A SEM image of synthesized copper oxide sensor.

598367 Application of Biophysical and Chemical Engineering Principles for Understanding Molecular Scale Interactions Critical to Virus Entry and Infection of Its Host

Abstract:

The coronavirus disease 2019 (COVID-19) pandemic has focused attention on the need to develop effective therapies against the causative agent, SARS-CoV-2, and also against other pathogenic coronaviruses (CoV) that have emerged in the past or might appear in future. Focusing on steps in the CoV replication cycle, in particular the entry steps involving membrane fusion that are vulnerable to inhibition by broad-spectrum or specific antiviral agents, is an astute choice because of the conserved nature of the fusion machinery and mechanism across the CoV family. For coronavirus, entry into a host cell is mediated by a single glycoprotein protruding from its membrane envelope, called spike (S). Within S, the region that directly interacts with the membrane is called the fusion peptide, FP. It is the physico-chemical interactions of the FP with the host membrane that anchors it, thus enabling the necessary deformations of the membrane that lead to delivery of the viral genome into the cell when a fusion pore opens. As chemical engineers, we contribute to this fundamental work by leveraging our understanding of thermodynamics, kinetics, and intermolecular interactions to describe FP interactions with the host membrane at the most fundamental molecular level to facilitate the development of strategies to limit those interactions to stop the spread of infection. In this talk, I will describe our work on understanding the impact of calcium ions on CoV infection. Using cell infectivity, biophysical assays, and spectroscopic methods, we found that calcium ions serve to stabilize the fusion peptide structure during conformational change that then allows its insertion into the host membrane, resulting in increased lipid ordering in the membrane. This lipid ordering precedes membrane fusion and has been shown to correlate with increased fusion activity, as higher extents of fusion are observed as calcium concentration increases, aligned with higher levels of infection in the presence of calcium. Finally, depletion of calcium ions leads to structure and activity changes that correlate well with *in vitro* experiments using calcium-chelating drugs. Under these conditions, cell infection dropped, pointing to the possibility of such drugs as therapeutic interventions.

612711 Portable Paper-based Biosensors and Novel Artificial Olfaction Biosensors Driven by Cell-free Synthetic Biology

Abstract:

Chemical or biological pollution in rapidly urbanizing cities has become a serious problem that disturbs the lives and livelihoods of millions of people, damaging and restricting potable water supply. Therefore, the rapid detection of pollution molecules will play a significant role in health care as well as environmental management. Many attempts, such as the utilization of diverse biological sensing elements, have been designed to respond to polluted chemicals or bacterial biomarkers for the detection. However, the practical applications of the whole-cell- or cell-element-based biosensors has been restricted by several concerns of biosafety, the hard preparation, and tough storage. For these challenges, cell-free protein synthesis technology, which synthesizes proteins using biological machinery *in vitro*, could be employed to broaden the applications of synthetic-biology-based biosensors or functional-protein-based biosensors.

For synthetic-biology-based biosensors, we present a portable *in vitro* biosensing platform for detecting inorganic chemicals or biological organic molecules. The sensing synthetic genetic circuits, coupled with basic cell-free components, were embedded onto papers by freeze-drying, which could be stored and transported at room temperature. These portable testing papers were activated by adding liquid samples for specific inducible reporter expression. To achieve the full availability outside the laboratory, reporter enzymes that could mediate color change by cleaving substrate were selected and demonstrated for *in vitro* paper-based platform. Synthetic genetic circuits respectively responded to two chemicals arsenic ion and AHLs, which were the common industrial pollutant and bacterial communication signal, were demonstrated to effectively sensing in this *in vitro* paper-based platform. Besides, combining the smartphone and image analysis software, the visible output could be easily collected and quantified. Without the intact cells, cell-free systems provide an emerging possibility for synthetic biology-based biosensor, which can be practically applied to environmental management or clinical diagnosis outside the laboratory. In the future, more internet technology such as smartphone and internet-of-thing (IoT) technology could be integrated into the detection platform for data sharing, from which people can learn about the environmental condition from a smartphone with the internet.

Moreover, the controllable protein synthesis conditions of cell-free systems also provide an emerging alternative approach for the effective production of complex proteins, such as the membrane olfactory receptor proteins. The nanomaterials can be applied for the functional folding of proteins in cell-free protein synthesis. Combining with the quartz crystal microbalance platform, diverse natural olfactory receptor proteins can be developed into artificial olfaction, which can be potentially employed to detect toxic odors for environmental management and applications.

611124 Biosensing of Human Hormone Disruptors in Blood and Urine with Protein Folding Using Cell-Free Protein Synthesis

Abstract:

The EPA has listed thousands of chemicals as potential hormone disruptors, however testing these chemicals is currently a lengthy and expensive process. Here we present a colormetric assay to rapidly detect estrogen or thyroid hormone activity of a sample using cell-free protein synthesis. The biosensor is based on protein folding of an allosteric fusion protein that includes a human thyroid or estrogen receptor. When a chemical binds to the hormone receptor during folding, the protein's confirmation changes which can then be detected by the reporter enzyme portion of the fusion protein. Using this approach, known hormone disruptors have been rapidly detected including the weakly binding estrogen disruptor BPA. Using lyophilized cell-free reagents, the sensor can be used in a "just-add-sample" format and has potential as a portable at-home assay. We also report the ability of this sensor to function in human blood and urine and the importance of adding RNAse inhibitors for sensing in these mediums.

611687 Fluorescent Microbial Sensors That Discriminate between Different Radionuclides.

Abstract:

Proper detection and monitoring of nuclear fuel cycle, enrichment, and weapon development activities is critical for supporting warfighter preparation in chemical, biological, radiological, nuclear, and explosives (CBRNE) operations, clandestine activities, and nuclear compliance. Field applications of conventional radiation monitoring and detection systems are restricted by several disadvantages: mandatory proximal placement near a radioactive source for detection, being easily identifiable and avoided, and capacity to only report radioactivity at a specific moment in time. The ability to detect trace amounts of ionizing radiation is a paramount concern for CBRNE operations and observation of clandestine activities in the event of source relocation and contamination. Microbes respond to environmental stress such as low-dose ionizing radiation with morphological and behavioral changes. Responses are known to persist after the stressor is removed. We hypothesize that transcriptional changes can monitor these stress responses and be harnessed to develop biosensors capable of discerning radionuclide type and monitor and report on nuclear fuel cycle, enrichment, and weapon development activities in diverse environmental conditions.

Unique responses of environmental microbes to post exposure to nuclear fuel cycle representative radionuclides were identified through mRNA sequencing analysis. Gene expression analysis showed *Pseudomonas putida and Escherichia coli* responded differently to acute and chronic exposure to plutonium-239, tritium, and iron-55 at a dose rate of 8.7 mGy/d. Highly abundant genes overexpressed by a sole radionuclide were selected for sensor development. Native promoter sequences from candidate genes were inserted into a broad host range plasmid upstream of a fluorescent reporter. Sequence verified *in vivo* sensors were then transformed into *P. putida* and subsequently irradiated by each of the three radionuclides for 24 hours before fluorescence was measured. Preliminary testing revealed, when grown in the presence of the inducing radionuclide, expression of GFP increased. Sensors treated with non-inducing radionuclides showed no significant increase in fluorescence. Future sensor development will include engineering circuits of toehold switch sensors permitting detection of time dependent signals for individual radionuclide sources. In addition to developing passive and autonomous monitoring or clandestine activities, the results of this work will expand upon the limited knowledge of low dose radiation effects in microorganisms.

599613 MRI/S Based Assessment of Donor Efficacy in Stroke Treatment with Human Mesenchymal Stem Cells (Industry Candidate)

Abstract:

Human mesenchymal stem cells (hMSC) have become a potential candidate for cell therapy in stroke treatment due to their strong secretion of trophic factors, immunomodulation and linage-specific differentiation. However, *in vitro* culture expansion leads to cellular senescence and reduced therapeutic functions after transplantation of 2D cultured hMSC, which remains one of the major barriers for biomanufacturing of clinical grade hMSC for stroke patients. To preserve and enhance stem cell properties during hMSC culture, preconditioning strategies such as 3D aggregation and hypoxia culture have been studied widely and show promising results both for *in vitro* and *vivo* hMSC efficacy. However, preclinical studies with preconditioned hMSC for stroke treatment still can provide inconsistent outcomes due to donor age and morbidity. For instance, it is challenging to determine the quality of donors with diseases or from an aged population by efficient *in vitro* cellular assays without long-term expansion, as most clinical procedures mandate short duration culture and limited modifications of cells prior to implant. Magnetic resonance imaging (MRI) provides excellent soft tissue contrast and is commonly applied to stroke diagnosis in both clinical and preclinical studies. However, MR techniques also can provide feedback on the efficacy of stroke therapies applied after the initial diagnosis.

In this study, treatment efficacy between two hMSC donors is investigated in a rat model of transient ischemia created by middle cerebral artery occlusion (MCAO). First, several key *in vitro* characteristics of hMSC with 3D aggregation preconditioning were analyzed. Second, to monitor the *in vivo* therapeutic effect from transplanted hMSC (health donor, compromised donor versus saline-injected immediately after MCAO), biochemical markers were measured longitudinally over 21 d using sodium chemical shift imaging (^{23}Na CSI), proton relaxation-enhanced MR spectroscopy (RE-MRS) and T_2-weighted proton imaging. Applying these techniques at the ultra-high field of 21.1 T provided increased sensitivity, enabling insight into ionic and metabolic homeostasis and displaying the dependence of tissue recovery on donor metrics.

In vitro results showed that hMSC from different donors have similar senescence, proliferation, migration and immunomodulation at passage 5, which is expansion point used for transplantation in MCAO rats. The functional decline in these assays for the compromised donor only can be observed beyond passage 7. *In vivo* results indicate ischemic lesion volume, as determined by ^{23}Na CSI, is significantly reduced by day 3 for the healthy donor hMSC cohort compared to the compromised donor and saline-injected MCAO controls, which both display expansion of the lesion. Percent change in both ^{23}Na and ^{1}H absolute signal within the ischemic lesion also is significantly reduced by day 3 for the healthy donor compared to the compromised donor and saline MCAO controls. Furthermore, widespread systemic effects are measurable as early as 24 h post-transplantation for ^{23}Na signal levels in the contralateral hemisphere, with significant elevation seen for the compromised and saline cohorts compared to the healthy donor, which displays a longitudinally stable and lower sodium signal. RE-MRS results support the ^{23}Na CSI analysis with reduced lactate-to-choline levels by day 3 and increased N-acetyl-aspartate and creatine ratios at day 1 for the healthy donor cohort only. In summary, *in vivo* MRI analysis demonstrates the possibility to evaluate donor efficacy of hMSC in preclinical studies and remedy inaccurate, low passage *in vitro* assessments based on traditional stem cell properties.

This work is supported by NIH (R01 NS102395).

602137 Engineered CRISPR/Cas12a System As a Sensitive Diagnostic Tool for Detecting HCV, HIV, and SARS-CoV-2

Abstract:

CRISPR/Cas (Cluster of Regularly Interspaced Short Palindromic Repeats/CRISPR-associated protein) systems have emerged not only as a precise gene-editing tool but also as a new sensitive diagnostic tool due to additional functionality of some newly discovered Cas proteins. Class II effector proteins, such as Cas12a, are considered as a better alternative to the CRISPR/Cas9 system due to their compact size, specificity, and efficiency. Besides executing successful gene editing, once Cas12a forms a ternary complex with guide RNA (crRNA) and target dsDNA and executes a double-stranded break (referred to as cis-cleavage), it executes nonspecific cleavage of ssDNA. This attribute, known as trans-cleavage, is only activated once bound to a target that has complementary base-pairing to the guide crRNA. When combined with a FRET-based reporter, a fluorophore connected to a quencher via a short oligonucleotide, the presence of the target activator can be validated. Such a phenomenon has been efficiently harnessed by SHERLOCK (Specific High-sensitivity Enzymatic Reporter unLOCKing) and DETECTR (DNA Endonuclease Targeted CRISPR Trans Reporter) to reliably detect nucleic acids.

Though rising as a versatile diagnostic tool that can detect nucleic acids with a high level of precision and accuracy based on complementary base pairing principle, current CRISPR/Cas12a systems are limited to a nanomolar detection limit without an amplification step. We, therefore, sought to overcome this limitation by finding ways to engineer CRISPR/Cas systems to enhance their sensitivity for nucleic acid detection. We extended the 3'- or 5'-ends of the crRNA with different lengths of DNA, RNA, and phosphorothioate DNA. Through selective engineering and design, we achieved amplified Cas12a trans-cleavage activity as high as 3.5-fold with our modified crRNA compared to the wild type crRNA on the target dsDNA eGFP fragment. Employing this phenomenon, we optimized a fluorescence-based detection assay that enabled us to detect nucleic acids without an amplification step with unprecedented sensitivity and with a limit of detection in femtomolar scale. This engineered crRNA-Cas12a was successfully applied towards detecting clinically relevant targets such as HCV, HIV, and SARS-CoV-2.

When combined with an isothermal target pre-amplification step such as RPA/LAMP, we observed an enhance in sensitivity down to a few copies of nucleic acids with our engineered crRNA-Cas12a system. Moreover, we combined this platform with lateral flow assay to create a rapid, portable, and inexpensive point-of-care paper strip tests that can be used by nonprofessionals. As a proof of concept, the paper strip test successfully displayed visibly positive signal when targeting synthetic RNA of SARS-CoV-2 down to 3 copies. Furthermore, such engineered point-of-care diagnostics could be extended to various other infectious diseases.

597876 CRISPR-Cas12a Mediated Universal Electrochemical Biosensing Platform (Faculty/Industry Candidate)

Abstract:

An accurate, rapid, and cost-effective biosensor for the quantification of disease biomarkers is vital for the development of early-diagnostic point-of-care systems. The recent discovery of the trans-cleavage property of CRISPR type V effectors makes CRISPR a potential high-accuracy bio-recognition tool. Herein, a CRISPR-Cas12a (cpf1) based electrochemical biosensor (E-CRISPR) is demonstrated, which is more cost-effective and portable than optical-transduction-based biosensors. Through optimizing the in vitro trans-cleavage activity of Cas12a, E-CRIPSR was used to detect viral nucleic acids, including human papillomavirus 16 (HPV-16) and parvovirus B19 (PB-19), with a picomolar sensitivity. An aptamer-based E-CRISPR cascade was further designed for the detection of transforming growth factor β1 (TGF-β1) protein in clinical samples. As demonstrated, E-CRISPR could enable the development of portable, accurate, and cost-effective point-of-care diagnostic systems.

613020 Building Accessible Hands-on Biology Classroom Activities with Sensors and Freeze-Dried Cell-Free Technology (Invited Speaker)

Abstract:

Hands-on demonstrations greatly enhance the teaching of STEM concepts and foster engagement and exploration in the sciences. While numerous chemistry and physics classroom activities exist, few biology labs are practical and accessible due to the challenges and concerns of growing live cells in classrooms. Here we introduce a platform to develop hands-on molecular and synthetic biology educational activities based on easy-to-use, shelf-stable, freeze-dried, cell-free (FD-CF) reactions, which are simply activated by water. By using DNA which encodes fluorescent proteins to create visual outputs with these FD-CF reactions, we created a variety of engaging modules to teach gene expression, cellular functions, and other basic molecular biology topics that are otherwise difficult to easily teach in a hands-on manner in a classroom. By expanding the platform to incorporate sensor-based components, such as RNA switches, we also developed modules that can teach interdisciplinary biology topics, including biochemistry, biomaterials, and synthetic biology, as well as basic laboratory skills such as pipetting, experimental design, and the scientific method. Our initial kit launched from this platform has already been used in hundreds of classrooms/programs across the country, and initial results and feedback suggest that the activities are accessible, easy to use, educational, and engaging for middle school, high school, and undergraduate students. Overall, the platform introduces low-cost, user-friendly, and hands-on activities that can be used in classrooms to improve the quality of biology education and open the door for student-driven, independent explorations in the life sciences.

612158 Big Data-Based Fault Detection with Advanced Analytics in the Pharma Industry

Abstract:

Motivated by market pressure to reduce product costs, regulatory incentive to modernise manufacturing and a growing environmental concern to reduce waste has led the pharmaceutical industry to improve its manufacturing process. The impetus to adapt and modernise the manufacturing process can largely be attributed to the promotion of the Quality by Design paradigm (QbD) and guidance on Process Analytical Technology (PAT) by the FDA in 2004. The promotion of these methods coincided with a time where the cost of computing and data storage had declined, but also with developments in non-invasive online sensor technology, such as Raman and Near-Infrared (NIR) spectroscopy. This culminated into an increase intro the application of multivariate analysis (MVA) and specifically Multivariate Statistical Process Control (MSPC).

MSPC exploits the Big Data generated from a complex chemical process such as pharmaceutical production and can be used for process understanding, troubleshooting and on-line fault detection. MSPC tools include multivariant process charts and soft sensors and can be combined with APC methods to improve overall plant efficiency. MSPC aims to use big data to better understand a process, to develop a definition of normal operation and then use that to detect faults on-line. There are three main categories to classify as SPC methods (1) model-based (2) knowledge-based (3) data-based. Data-based methods are often the preferred method for pharmaceutical manufacturing. The advantages of data-driven methods are that they do not require first principals understanding of the process or a unit as they construct a model using input and output data. The most popular data-driven methods are multivariate latent/MVA methods (which include PCA, PLS and their adapted versions). These methods project data into the latent space and work as a feature extraction method and reduce the dataset into a smaller or manageable set that can be used in further applications of fault detection.

This paper presents an industrial case study which focused on fault detection towards operational improvements for a key API production, pursued in collaboration with a major multinational UK pharmaceutical company (GSK). The aim of this project is to exploit data availability towards constructing and validating a data-driven model of a specific API production unit of the GSK plant (Montrose, Scotland, UK). The project involves using large data sets of process data (temperatures, partial pressures, flow rates, and stream compositions, from an on-site GCMS analysis station), and implementing advanced MVA methods in MATLAB, in order to diagnose operational patterns and recommend feasible improvements towards reliable API production intensification.

LITERATURE REFERENCES

[1] E. Tomba et al., “General procedure to aid the development of continuous pharmaceutical processes using multivariate statistical modeling-An industrial case study,” Int. J. Pharm., vol. 444, no. 1–2, pp. 25–39, 2013.

[2] T. Kourti, “Application of latent variable methods to process control and multivariate statistical process control in industry,” Int. J. Adapt. Control Signal Process., vol. 19, no. 4, pp. 213–246, 2005.

[3] J. F. MacGregor and T. Kourti, “Statistical process control of multivariate processes,” Control Eng. Pract., vol. 3, no. 3, pp. 403–414, 1995.

[4] T. Kourti, “Multivariate dynamic data modeling for analysis and statistical process control of batch processes, start-ups and grade transitions,” J. Chemom., vol. 17, no. 1, pp. 93–109, 2003.

[5] J. F. MacGregor, P. Nomikos, and T. Kourti, “Multivariate Statistical Process Control of Batch Processes Using PCA and PLS,” IFAC Proc. Vol., vol. 27, no. 2, pp. 523–528, 1994.

[6] S. Garcia-Munoz and D. Settell, “Application of multivariate latent variable modeling to pilot-scale spray drying monitoring and fault detection: Monitoring with fundamental knowledge,” Comput. Chem. Eng., vol. 33, no. 12, pp. 2106–2110, 2009.

[7] M. Boiret, D. N. Rutledge, N. Gorretta, Y. M. Ginot, and J. M. Roger, “Application of independent component analysis on Raman images of a pharmaceutical drug product: Pure spectra determination and spatial distribution of constituents,” J. Pharm. Biomed. Anal., vol. 90, pp. 78–84, 2014.

598331 Recent Advances in Kinetic Parameter Estimation Toolkit (KIPET) with Spectra

Abstract:

Developed over the past four years, the Kinetic Parameter Estimation Toolkit (KIPET), is an open-source toolbox for the determination of kinetic parameters from a variety of experimental datasets including spectra and concentrations. *KIPET* seeks to overcome limitations of standard parameter estimation packages by applying a unified optimization framework based on maximum likelihood principles and large-scale nonlinear programming strategies for solving estimation problems with nonlinear differential algebraic equations (DAEs). The software package includes tools for data preprocessing, determination of parameter confidence levels for a variety of problem types, and informative wavelength selection to improve the lack of fit (Schenk et al., 2020). All these features have been implemented in Python with the algebraic modeling package *Pyomo*. *KIPET* exploits the flexibility of *Pyomo t*o formulate and discretize the dynamic optimization problems that arise in the parameter estimation algorithms. Solutions of these optimization problems are obtained with the nonlinear solver *IPOPT* and confidence intervals are obtained through the use of either *sIPOPT* or a newly developed tool, *k_aug*.

In this talk, we discuss a recently enhanced *KIPET* package (Short et al., 2020) that considers multiple experiments with potentially different reactants and kinetic models, different dataset sizes with shared or unshared individual species' spectra, leading to fast parameter estimation and confidence intervals based on the NLP sensitivities. In addition, we present a new variance estimation technique based on maximum likelihood derivations for unknown covariances from two sample populations. This approach leads to a straightforward deconvolution of variances between noise in model variables and in spectral measurements. Moreover, we discuss a new estimability analysis approach that systematically determines a ranked subset of kinetic parameters with well-defined confidence intervals (Chen and Biegler, 2020). With nonlinear kinetic models and limited measurements, it is often difficult to correctly estimate all the parameters, due to linear dependence and low correlation among the parameters. A common approach is to estimate a subset of the parameters by fixing the others at reasonable (often literature) values. However, it is often challenging to determine which parameters can be properly estimated. In this talk we present an efficient approach that ranks the estimable parameters, and discards those that cannot be estimated accurately. Based on reduced the reduced Hessian information with simple Gauss-Jordan elimination, the proposed approach leads to fast parameter selection and estimation within a simultaneous collocation framework. This approach becomes much more efficient for large problems than competing approaches based on multiple eigenvalue decompositions (Quaiser and Mönnigmann, 2009). Several case studies with increasing complexity are presented to demonstrate the performance of this proposed approach.

1. W. Chen, L. Biegler, “Reduced Hessian Based Parameter Selection and Estimation with Simultaneous Collocation Approach,” submitted for publication (2020)
2. Schenk, M. Short, J. S. Rodriguez, D. Thierry, L. T. Biegler, S. Garcia-Munoz, W. Chen, “Introducing KIPET: A novel open-source software package for kinetic parameter estimation from experimental datasets including spectra," Computers and Chemical Engineering, 134(4), 106716 (2020)
3. Short, L. T. Biegler, S. Garcia-Munoz, W. Chen, "Estimating Variances and Kinetic Parameters from Spectra Across Multiple Datasets Using KIPET," Chemometrics and Intelligent Laboratory Systems, to appear (2020)
4. Quaiser, T., Mönnigmann, M. “Systematic identifiability testing for unambiguous mechanistic modeling-application to JAK-STAT, MAP kinase, and NF-κB signaling pathway models,” BMC Systems Biology, 3,50 (2009).

601039 Teaching Computers to Interpret Mdsc Thermograms

Abstract:

We present a novel algorithm for the automatic data analysis of modulated Differential Scanning Calorimetry of thermograms designed to minimize subject matter expert input reduce the man reliability.

Differential Scanning Calorimetry (DSC) and mainly modulated DSC (mDSC) analysis are widely used to support the production of amorphous solid dispersions for the identification of kinetic and thermodynamic events such as glass transition, crystallization and melting. However, interpretation of thermograms require careful analysis of subject matter experts (SMEs) which can be time-consuming and thus can be a bottleneck to, e.g., formulation development where large numbers of samples are generated are require prompt analysis for reiteration. Furthermore, often there are elusive phenomena that can sometimes go unnoticed or cause dubious interpretation from different SMEs.

The presented Python-based software tool's computation of the derivatives and the maximums, minimums and inflections of each curve, is the base of the software. It then uses this information to search for events, based on their magnitude. One essential advantage of using an automated software is the ability to use the derivatives of the heat flow curves to improve the precision of the results. The derivatives are essential not only to the detect the exact temperature at which events occur but also to help define their onsets. In order to tackle the inherent noise of the DSC analysis and data acquisition, the proposed tool employs a low pass, Fourier transform-based filter, which enables the use of the thermogram derivatives.

We show case studies where the software analysis increases the detection rate of events that can easily be unnoticed by visual assessment due to their magnitude or shape, as they only cause very subtle variations. We also present case studies where smaller events are distorted or confounded with a nearby dominant events. However, following the digital detection of these events a more detailed visual assessment confirms their presence and often indicative of the onset of undesirable events such as phase separation or crystallization. Furthermore, the presented digital analysis tool promotes a faster and objective analysis of generated DSC thermogram data.

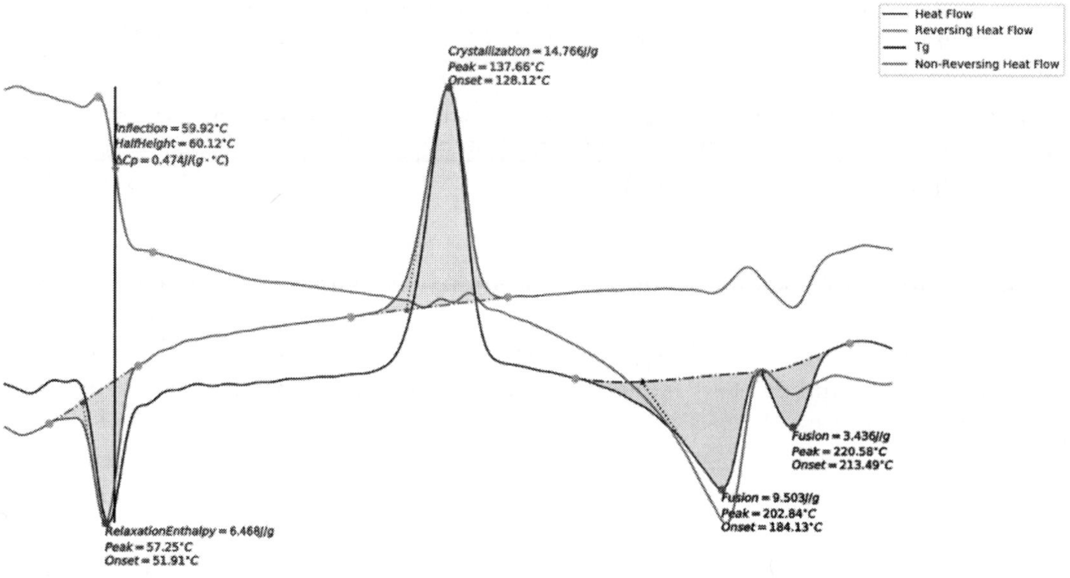

595673 Data-Smart Machine Learning Methods for Predicting Youngs Modulus of Directly Compressed Blends of Pharmaceutical Powders

Abstract:

The ability to predict the mechanical properties of compacted powder blends of Active Pharmaceutical Ingredients (API) and excipients solely from those of the individual components can reduce the amount of "trial-and-error" involved in formulation design. Machine Learning (ML) models can reduce model development time and effort with the imperative of adequate historical data. This work evaluates linear and non-linear ML for predicting the Youngs Modulus of directly compressed arbitrary blends of known excipients and API from readily available information about the API. The training data obtained from three BCS Class I APIs and four excipients demonstrate "data-smart" strategies to train ML models efficiently. The results indicate that even simple linear ML model provide reasonable prediction accuracy. The practical benefits of this method and how it compares with other mechanistic models are discussed. Finally, we demonstrate an application of the model to enable Quality-by-Design in drug product development.

593356 Advances and New Developments on Eiot for PAT Applications in Pharmaceutical Development and Manufacturing

Abstract:

The EIOT method[1] was proposed as an extension to the IOT[2] method as an alternative approach to extracting information from spectroscopic data into interpretable mass fractions of the resolved chemical species in the powder mixture.

The method in essence seeks to use Beer-Lambert's law as the deterministic functional relationship between spectra and mass fractions. The main difference with IOT is that EIOT does not assume the spectrum of the pure components is a measurable quantity; but a quantity that can only be estimated from measurements of known composition. The EIOT method proposes to decompose the measured spectra as the summation of the chemical signal plus the effect of the non-chemical interferences. The chemical signal is given by the summation of the apparent pure spectrum weighted by the mass fraction of each specie; the non-chemical interferences as the summation of the non-chemical interference signals weighed by their respective strength. The method relies on optimization techniques to estimate the apparent pure spectra and the signal of the non-chemical interferences; and when the method is used in real-time, to estimate the mass fractions and the strengths of the non-chemical interferences in each new sample.

Since the method was published, a number of enhancements have been done to the method, this talk presents such enhancements that i) the consideration of supervised signatures and ii) the consideration of exclusivity in a set of signatures (i.e. only one in a set is allowed to contribute). Details around the maintenance and long-term ownership of the EIOT method are discussed for its potential inclusion as a central component for a control strategy in drug product manufacture.

EIOT (as well as its predecessor IOT) bring a disruptive change in the practice of chemometrics; retiring the use of fixed regression models, to the use of real-time optimization and computer decision-making instead. As such, the requirements and expectations for use and long-term maintenance might have to be revised and adapted.

[1] Z. Shi, J. Hermiller, S.G.J.A.J. Muñoz, Estimation of mass-based composition in powder mixtures using Extended Iterative Optimization Technology (EIOT), (2018).

[2] K. Muteki, D.O. Blackwood, B. Maranzano, Y. Zhou, Y.A. Liu, K.R. Leeman, G.L. Reid, Mixture component prediction using iterative optimization technology (calibration-free/minimum approach), Industrial & Engineering Chemistry Research, 52 (2013) 12258-12268.

602264 Modeling and Nonlinear State Estimation for Advanced Process Control of the Enzymatic Conversion of Lactose into Value-Added Products

Abstract:

Due to increased concerns over sustainability of chemical processes, significant research has been conducted to minimize the amount of process waste generated. Despite advances in efficiency, selectivity, and conversion, waste management is still a concern for all processes, especially biological. The most effective method for waste disposal would be its conversion into a value-added product through profitable means. Many industries are attempting to achieve this goal, but it is challenging to design a novel catalyst, enzyme, or process. In particular, recent advances in enzymatic process modeling have created an opportunity for small-scale cheese manufactures to convert their waste into value-added products. The manufacturing of cheese results in the formation of lactose rich whey permeate that cannot be easily utilized and thus is often disposed of at cost to manufacturers [1].

A recent advance in enzymatic modeling corresponds to a new kinetic model to convert lactose waste into value-added galactooligosaccharide (GOS) products using β-galactosidase [2]. This specific reaction has been studied by several researchers in the past, but each proposed model presented challenges such as mass balance issues and overfitting due to the excessive number of parameters [3]. The proposed kinetic model contains no mass balance errors, while minimizing the number of parameters to accurately depict the process characteristics. Furthermore, this model uniquely distinguishes between purely galactose derived disaccharides, trisaccharides, and tetrasaccharides and those containing a glucose saccharide group. Due to the similarity of glucose and galactose, there are no online measuring techniques that can discern between the specific di, tri, or tetrasaccharides stereoisomers. A High Performance Liquid Chromatography (HPLC) method coupled with an ion-exchange column and refractive index detector can distinguish between glucose and galactose, but has no resolution for other isomers such as disaccharide, trisaccharide, and tetrasaccharide stereoisomers. Without the ability to measure specific component concentrations, state estimation must be used to predict the individual states using the online measurements and the proposed kinetic model.

This work addresses the state estimation and advanced control employing the developed enzymatic model for the first time. Unique estimation challenges are posed and addressed, with the specific novelty of using nonlinear state estimation to find stereoisomer concentrations from available online measurements and the nonlinear kinetic model. Both an Extended Kalman Filter (EKF) and a Moving Horizon Estimator (MHE) are analyzed for process implementation. The EKF is an unconstrained estimator that allows for rapid estimation but does not guarantee feasible and optimal results. The MHE is a constrained estimator that guarantees feasibility at the cost of increased computational time [4] [5]. Using these estimation techniques, a Biologically-Inspired Optimal Control Strategy (BIO-CS) algorithm is applied to optimize the feeding of lactose and enzyme into a semi-batch reactor [6][7]. The proposed estimation and advanced control framework allows the feasibility of this process to be examined and provides a guide for other bioprocesses to follow.

References:

[1] Illanes, A. (2011). Whey Upgrading by Enzyme Biocatalysis. *Electronic Journal of Biotechnology*.

[2] Schultz, G., Alexander, R., Lima, F.V., Giordano, R., & Ribeiro, M. (2019). Kinetic Model for the Enzymatic Synthesis of Galacto-Oligosaccharides: Describing Galactobiose Formation. *AIChE Annual Meeting.*

[3] Vera, C., Guerrero, C., Illanes, A., & Conejeros, R. (2011). A Pseudo Steady-State Model for Galacto-Oligosaccharides Synthesis With ß-Galactosidase From *Aspergillus oryzae. Biotechnology and Bioengineering.*2270-2279.

[4] Campani, G., Ribeiro, M., Zangirolami, T., Lima, F.V. (2019). A Hierarchical State Estimation and Control Framework for Monitoring and Dissolved Oxygen Regulation in Bioprocesses. *Bioprocess and Biosystems Engineering.*1467-1481.

[5] Lima, F.V., Rawlings, J. (2011). Nonlinear Stochastic Modeling to Improve State Estimation in Process Monitoring and Control. *Process Systems Engineering.*996-1007.

[6] Mirlekar, G., Gebreslassie, B., Diwekar, U., Lima, F.V. (2018). Biomimetic Model-Based Advanced Control Strategy Integrated with Multi-Agent Optimization for Nonlinear Chemical Processes. *Chemical Engineering Research and Design.*229-240.

[7] Mirlekar, G., Al-Sinbol, G., Perhinschi, M., Lima, F.V. (2018). A Biologically-Inspired Approach for Adaptive Control of Advanced Energy Systems. *Computers and Chemical Engineering.*378-390.

604108 Designing Data-Based and Model-Based Methods for Process Monitoring and Equipment Degradation Tracking

Abstract:

A wide variety of industrial processes utilize monitoring methods in order to ensure that safety and quality is maintained. Conventional model-based methods include dimensionality reduction techniques such as Principal Component Analysis (PCA) are often employed as they are computationally simple, and easy to implement in practice [1]. Unfortunately, these techniques assume that the process data is Gaussian, uncorrelated, and only contain a moderate level of noise. This work will demonstrate how wavelet-based representation of data can be utilized in order to efficiently handle data that violate these assumptions. Statistical hypothesis testing methods such as the generalized likelihood ratio chart have shown promise with respect to fault detection, and thus will be used to enhance monitoring capabilities [2], [3].

Additionally, this work will demonstrate the necessity to track deviations in the process models themselves in order to identify process drifts or equipment degradation. When a process operates under control, a controller continuously adjusts the level of manipulated variables in order to ensure that the controlled variables are being maintained within specific limits. Unfortunately, this may be at the expense of increased operating and additional costs. Therefore, this work will utilize a dynamic contour-based algorithm in order to illustrate how equipment degradation can be efficiently tracked in multiple operating regimes [4].

For both algorithms, illustrative examples using simulated synthetic data, and real data from different applications will be utilized in order to highlight the effectiveness of the developed algorithms.

References

[1] I. T. Joliffe, *Principal Component Analysis*, 2nd ed. New York, NY: Springer-Verlag, 2002.

[2] M. R. Reynolds and J. Y. Lou, "An Evaluation of a GLR Control Chart for Monitoring the Process Mean," *J. Qual. Technol.*, vol. 42, no. 3, pp. 287–310, 2010.

[3] M. Z. Sheriff, M. Mansouri, M. N. Karim, H. Nounou, and M. Nounou, "Fault detection using multiscale PCA-based moving window GLRT," *J. Process Control*, vol. 54, 2017, doi: 10.1016/j.jprocont.2017.03.004.

[4] M. Z. Sheriff, H. Nounou, M. Nounou, and M. N. Karim, "Monitoring process degradation through operating regime based process monitoring," in *AIChE Spring Meeting and Global Congress on Process Safety: Process Control Monitoring and Analytics*, 2019.

610824 Optimal Sensor Network Design and Nonlinear State Estimation for in-Situ Corrosion Monitoring in Coal-Fired Boilers

Abstract:

Existing coal fired power plants are increasingly being subjected to significant load-following due to the increased penetration of intermittent renewables to the grid. During load following operations maintaining the boiler environment at the desired conditions is significantly difficult. In particular, higher flue gas temperature than expected and higher concentration of corrosive sulfur species in the flue gas are causing external sulfidic corrosion especially in the waterwall (WW) tubes, which is one of the primary causes for the WW failure [1]. Experimental studies on corrosion rates of Ni-Cr alloys suggest that corrosion rate is a function of alloy composition, temperature, SO_2 and O_2 concentrations in the flue gas [2-3]. The effect of metal temperature on the corrosion rate is highly nonlinear with a "bell" shaped response between 600°C and 1000°C for most of the alloys used in the WW section. Unfortunately, as of now, there is no online commercial sensor for real-time measurement of corrosion rate due to the high operating temperature and challenging environment in the WW section. Recently a novel type of electrochemical sensor is being developed in-house for in-situ measurement of the corrosion rate [4]. As the operating temperature and concentration of sulfur species greatly vary spatially and temporally in the WW, the corrosion rate can vary significantly with space and time. However, it is infeasible to install such sensors in the entire WW. Motivated by these challenges, an optimal sensor network is designed, and a nonlinear state estimation technique is developed for real-time monitoring of corrosion rate at all desired locations inside the WW.

A first-principles dynamic process model of the WW is developed for calculating the spatial and temporal variation in the temperature and concentration of sulfur species. Models are also developed for the corrosion rate. Electrochemical sensor model has been previously developed at West Virginia University [4]. This is a high nonlinear differential algebraic equation (DAE) system. Therefore, an unscented Kalman filter algorithm for DAE system is developed for state estimation, particularly for estimating the corrosion rate given the data from the electrochemical sensor.

Typically, process information which can be obtained from a given sensor network and the reliability of sensor network improves as the number and/or the accuracy of the installed sensor increases. However, it is neither feasible nor necessary to install sensors at all desired locations. An optimal sensor network is designed by minimizing the posterior error covariance of the optimal estimator. The algorithm provides the optimal number, location and type of sensors.

The optimal sensor network is used to estimate the corrosion rate for a coal-fired boiler. A number of operating scenarios are simulated by changing the temperature in the WW and concentration of the sulfur species. Our work shows that even in the presence of large mismatch between the process and the model and high measurement uncertainty, the sensor network provides satisfactory spatial and temporal resolution of the corrosion rate under aggressive load-following operation.

References

1. North American Electric Reliability Council (NERC), "State of Reliability," 2017.

o

2. L. Luthra and D. A. Shores, "Mechanism of Na_2SO_4 Induced Corrosion at 600-900 C," *J. Electrochem Soc.*, no. 1971, pp. 2202–2210, 1978.
3. Viswanathan and C. J. Spengler, "Corrosion of 85 Ni-15 Cr Alloy at 1600 F in Controlled Atmospheres Containing O_2, SO_2, SO_3, H_2S, and N_2," *Corrosion*, vol. 26, no. 1, pp. 29–41, 1970.
4. N. Aung and X. Liu, "High temperature electrochemical sensor for in situ monitoring of hot corrosion," *Corros. Sci.*, vol. 65, pp. 1–4, 2012.
5. Kumari, S. K. Das, and P. K. Srivastava, "Modeling fireside corrosion rate in a coal fired boiler using adaptive neural network formalism," *Port. Electrochim. Acta*, vol. 34, no. 1, pp. 23–38, 2016.

609445 CO_2 Injection Monitoring with a Scalable, Automated Sparse Seismic Array

Abstract:

Successful applications of the 3D surface land seismic method to monitor CO_2 injection into a reservoir have been found in the literature. Reducing the operational effort to acquire 3D data and the turnaround time for the processing and interpretation of field data are highly desirable for more efficient seismic monitoring. The scalable, automated sparse seismic array (SASSA) method has been ideated as a cost-effective alternative or complementary to conventional 3D seismic methods. Based on a sparse array of receivers deployed strategically on Earth's surface and one or more fixed seismic sources, the SASSA method can be used to study specific portions of the subsurface to detect the movement of CO2 in a reservoir with lower environmental footprint and shorter turnaround time to deliver results than conventional 3D seismic methods.

The Energy & Environmental Research Center (EERC) carried out seismic monitoring of an oilfield undergoing CO_2 enhanced oil recovery in Montana, USA, using the SASSA method. Initial efforts in collaboration with the field operator were invested in designing an optimum distribution of two sources and 96 receivers not only to honor the subsurface geologic conditions and CO_2 injection patterns but also to minimize the impact of cultural noise sources. The signal-to-noise ratio was further improved in data acquisition by use of two fixed surface orbital vibrators that were remotely controlled by the EERC team in North Dakota. Weekly data acquisition was conducted on weekends from May 2019 through March 2020. Data-processing workflows were developed to preserve the true seismic amplitudes representing the time-lapse changes of the reservoir due to CO_2 injection and maximize the representation of the reservoir in the seismic sections from the distribution of sources and receivers on the surface. Encouraging results were obtained using the SASSA method in areas adjacent to the CO_2 injection wells. Two validation methods were considered: a time-lapse 2D seismic survey acquired at the beginning and the end of the monitoring activities and reservoir simulation.

598416 (Invited Plenary Talk) Wearable Electrochemical Sensors

Abstract:

Wearable sensors have received a major recent attention owing to their considerable promise for monitoring the wearer's health and wellness. The medical interest for wearable systems arises from the need for monitoring patients over long periods of time. These devices have the potential to continuously collect vital health information from a person's body and provide this information to them or their healthcare provider in a timely fashion. Such sensing platforms provide new avenues to continuously and non invasively monitor individuals and can thus tender crucial real-time information regarding a wearer's health. Our research over the past decade has aimed toward filling the gaps toward providing biochemical information, beyond that given by common wrist-watch mobility trackers. This presentation will discuss recent developments in the field of wearable electrochemical sensors integrated directly on the epidermis or within the mouth for various non-invasive biomedical monitoring applications. Particular attention will be given to non-invasive and minimally-invasive monitoring of metabolites and electrolytes using skin-worn and microneedle amperometric and potentiometric sensors, respectively, along with related materials, energy and integration considerations. The preparation and characterization of such wearable electrochemical sensors will be described, along with their current status and future prospects and challenges.

610185 (Invited Plenary Talk) an Integrated Paradigm for Precision Exposure to Airborne Chemical and Biological Stressors Based on Personal Sensing

Abstract:

Introduction

The current manuscript outlines the fundamentals of a methodological framework for precision exposure assessment and gives concrete examples of its application in different European cities to assess the health risk associated with exposure to fine and ultra fine particles of chemical and biological origin. The work described herein involves a multi-city personal sensor campaign aiming to refine PM exposure using low-cost portable sensor data, exposure and human respiratory tract deposition modelling to estimate the biologically effective internal daily intake on an individual basis and agent-based modelling to estimate exposure and consequent health risk to the community.

Methodology

A prototype monitoring device (IcaruSense) was developed for measuring three size fractions of PM (1, 2.5 and 10 μm), enabling direct assessment of personal exposure. The device is based on an Arduino microcontroller where miniaturized electrochemical sensor-modules are connected. In order to improve sensor performance, correction factors were used to account for diverse environmental conditions and they were calibrated with the help of a well-validated light scattering optical sensor. Study participants wore a physical activity wristband (Garmin Vivosmart 3) that records steps, distance, type of activity, heartbeat and sleeping patterns. Finally, participant positions were recorded using a GPS sensor, integrated into the PM sensor kit. After calibration and validation, the IcaruSense devices were used to capture daily variability of PM exposure. Exposure was further refined by estimating inhalation adjusted intake rate from the continuous time-activity data collected from the physical activity and the GPS sensors. Using the inhalation adjusted intake rate PM deposition across the human respiratory tract (HRT) was readily estimated using the Multiple Path Particle Deposition (MPPD) model.

A city scale agent-based model (ABM) was developed to support this work by expanding the precision exposure estimates from the individual to the community level. The model feeds into population-based exposure assessment without imposing prior bias, basing its estimations onto emergent properties of the behaviour of the computerised autonomous decision makers (agents) that compose the city-system. Population statistics, road and buildings network data were transformed into human, road and building agents, respectively. Informed by literature-extracted relationships between Socio-Economic Status (SES) and human behavioural patterns, the established ABM simulates societal dynamics. Time-use survey outputs were associated with human agent behavioural rules, aiming to model representative to real-world routines. Moreover, real time-geography of exposure data, extracted from the precision sensors campaigns in nine European cities were used to parametrise and enhance the model.

Mobility behaviour and eventually exposure to pollutants change as a result of a human agent-specific decision making. Virtual individuals of different sociodemographic backgrounds use different means of transportation and follow different sequence and types of activities. Behaviours that were not explicitly programmed, arise through the human agents' interactions, enabling the examination of emergent behaviours from the bottom-up. Spatiotemporal trajectories are coupled with spatially resolved Particulate Matter (PM) levels, enabling the assessment of personal and inhalation adjusted exposure, based on type of location and intensity of enrolled activities.

Results and Discussion

The above methodology was applied in the personal sensors campaign of the HORIZON2020 EU Project ICARUS, where exposure and intake to PM of almost 100 individuals in each out of nine European cities were monitored. The integrated hybrid approach outlined above, which couples multi-sensor datasets on an individual basis with advanced respiratory tract modeling, allowed us to significantly differentiate the actual intake of the participants, highlighting larger differences than the ones attributed to the spatial differentiation of air pollution based on fixed stations (difference of ambient PM levels of 50% were translated in intake differences up to 110%). These differences are the result of the differences in PM size fractions that are captured by the sensors and the capability of the HRT model to translate these differences in PM size distribution accounting for variation in respiratory tract physiology among the participants, which is ultimately reflected in intake estimates.

ABM findings indicated that inhalation adjusted exposure differences between 2 individuals living in the same neighbourhood can vary by as much as 87%, due to radically different spatiotemporal behaviours. This major difference was observed for cases where one human agent has a full-time job and often exercises in outdoor courts whereas his/her neighbour is a retired homemaker. For the same reason, ABM runs confirmed that even people who reside in the same dwelling might not necessarily be exposed to similar levels of pollution. Model results indicated that PM inhalation adjusted exposure between flatmates can differ by 61%. Evidence was, also, provided for vulnerable subgroups of population that are disproportionately subjected to high levels of exposure. ABM-retrieved population exposure findings indicated that subgroup median exposure for children, adult males with a full-time job and low income and human agents of a lower educational level was notably higher than the total population median value. The validated model can assess exposure for the entire sociodemographic spectrum of a simulated region, without the constant need for repeated individual monitoring which is often not feasible due to high costs, ethical constraints and complex organisation that is associated with such measurements. By modelling the heterogeneous routines of human agents, the ABM produces detailed information related to the societal system examined and generates data that could be used to fill in the gaps that exist in traditional datasets.

Conclusions

This study presents a new methodological approach to estimate the external exposome, which encompasses the totality of human environmental exposures at an individual and community level. This method helps to integrate substantive considerations of individual status into long-term urban planning and management; it would also facilitate the identification of both physically and socially practical means for reducing life-threatening exposure levels. The methodological paradigm for precision assessment of exposure at the individual and community levels laid out in this work can be used for evaluating the impacts of different public health policies prior to implementation.

614024 (Invited Talk) Developing Antibody-Based Nanoplasmonic Barcode Biosensors for Sars-Cov-2 Detection

Abstract:

The emergence of COVID-19 as an international pandemic has caused unprecedented disruptions to modern life. As economies reopen and society adjusts to a new normal, there is a critical need to not just diagnose currently or previously infected individuals but also to be able to rapidly detect the presence of viral particles in the environment. As supported by an NSF RAPID grant, this talk will describe work the authors are doing to develop antibody-based sensor devices to meet this need. Three strategies are being pursued in parallel to identify viable pairs of antibodies: *in silico* design of novel binders, rational engineering of SARS-CoV binding antibodies, and *in vitro* screening of antibodies from the spleens of MERS-immunized mice. Once candidate antibodies are identified, they are subsequently assessed for their ability to detect SARS-CoV-2 in complex fluids via a sandwich assay in a nanoplasmonic barcode device.

614095 (Invited Talk) Development of Lateral Flow Tests for Sars-Cov-2 Antibody and Antigen Detection

Abstract:

As SARS-CoV-2 began to spread across the United States, it became clear that rapid and reliable tests for antigen and antibody were needed. The mission at NOWDiagnostics is to develop simple, accurate, and easy to administer diagnostic tests that yield results in minutes. In response to the pandemic, we quickly shifted focus to the development of antibody serology tests, antibody saliva tests, and antigen saliva tests using the existing AdexusDx platform. Here I will discuss some of the obstacles we faced, as well as the successes.

614156 (Invited Talk) Rapid and Reliable Detection of Sars-Cov-2 Using a Smartphone-Based Portable Isothermal Amplification Platform

Abstract:

SARS-CoV-2 – the viral pathogen responsible for the COVID-19 pandemic outbreak –has exposed some of the key limitations facing current infectious disease management strategies. A need, therefore, exists for inexpensive and robust tools that can be broadly deployed to accelerate diagnosis and provide real-time data to improve decision making. To address this, we propose to integrate smartphones with a portable molecular diagnostic platform (based on LAMP) to rapidly detect viral genetic material in the human samples using machine learning. We also incorporate a thermodynamic model that can significantly decrease the assay optimization time and rate of false positive of LAMP reactions.

614023 (Invited Talk) Quantitative and Ultra-Sensitive Saliva Test with Cellphone for COVID-19

Abstract:

More than six months after the initial COVID-19 outbreak, most countries are still struggling to control COVID-19 disease transmission, and many are now observing increased infection rates in response to attempts to reopen public venues to reduce economic and social disruption. Current RT-PCR-based COVID-19 assays have drawbacks that limit their capacity for screening efforts required to inform such decisions, since they require significant technical expertise and expensive equipment; use nasopharyngeal samples that must be collected by well-trained personnel wearing extensive protective equipment; and miss a significant number of COVID-19 cases. Other recently approved assays can require less expensive equipment and technical expertise, but exhibit reduced sensitivity and specificity. We recently reported the development of a rapid and sensitive COVID-19 assay in which CRISPR-mediated activation of a fluorescent probe amplifies RT-PCR signal for SARS-CoV-2 RNA to increase assay sensitivity using nasopharyngeal swab samples. We now report that this approach demonstrated similar analytical sensitivity when applied to analyze saliva samples spiked with SARS-CoV-2 RNA, and that results from clinical saliva samples analyzed by this approach exhibited complete concordance with RT-PCR results from paired nasopharyngeal swab samples. This assay approach was used to directly detect SARS-CoV-2 RNA in saliva, without a separate RNA isolation step, and was not dependent upon specific incubation temperatures. We next demonstrated that this assay could be performed using a procedure that did not require any laboratory equipment, using an inexpensive 3D-printed adaptor to allow accurate and sensitive detection of SARS-CoV-2 RNA in clinical saliva samples for COVID-19 diagnosis.

614097 (Invited Talk) COVID-19 Risk Emergency Toolbox - an Integrated System for Reliable Early Alert and Effective Public Health Risk Management

Abstract:

One of the key aspects of managing efficiently the public health risk of COVID-19 is the ability to assess the extent of the contagion spread in the community. In this paper we propose a combined approach integrating environmental surveillance data from a network of biosensors applied on sewage wastewater and fine/ultrafine air particles with human clinical data monitoring in potential patients or survivors of COVID-19 using a wearable device for telemedicine. These datasets are further fused with climatic data such as air temperature, humidity and UV radiation flux that may be linked to infectious diseases such as COVID-19. The integrated environmental/clinical dataset is then post-processed with machine learning algorithms to derive estimates of the degree of infection spread in the community at very early stages of infection, using the environmental parameters as sentinels for the presence of SARS-CoV-2 in the community. These estimates are then used as input to an already validated multi-stage SEIR-X model (CORE), which supports the management of the COVID-19 health risk by capturing reliably its dynamics, including the impact of risk reduction measures. We demonstrate the predictive capacity of the integrated CORE system in applications in Northern Italy and Greece even in the light of significant uncertainties with regard to the actual spread of the infection in the community due to limited molecular testing on humans. The integrated model is already part of the OECD COVID-19 computational toolbox.

612009 (Invited Plenary Talk): Towards Ubiquitous Physiological Monitoring

Abstract:

Modern medicine is largely reactionary. We wait for patients to exhibit symptoms and then we try to treat them. Yet we know physiological changes precede clinical deterioration. Two of the most basic and critical physiological parameters are respiration and hemodynamics. Even short interruptions of either have catastrophic consequences. Remarkably, there is currently no way to automatically and **continuously** assess respiratory effort, minute ventilation, or even blood pressure.

In response, we have developed Band-Aid© like sensors for continuous and quantitative monitoring of both respiratory and hemodynamic waveforms. Using commodity shrink film to create nanostructured thin metal films, we have developed a lift off process to transfer these thin films into any elastomeric polymer to create scalable sensors for physiological monitoring. From these waveforms, we can continuously monitor key parameters – and importantly, how these parameters change in real time – including respiratory rate, respiratory flow, heart rate, and blood pressure. Both patches can be applied and left on for > 24 hours, and has been correlated with FDA approved medical grade devices (spirometer and finger-cuff volume clamp beat to beat blood pressure monitor, respectively). The biggest advantage of our sensor system compared to other solutions (e.g. pulse oximetry, capnography, blood pressure cuff, EKG leads) is its ability to provide automatic and continuous untethered ambulatory monitoring to alert of early indicators of acute distress. Additionally, because our sensors are compatible with roll to roll mass manufacturing, they are extremely inexpensive to produce and can be disposable.

The data from the wireless sensors can be transmitted wirelessly to a tablet, phone, or workstation. With these sensors, we have various human subject tests in different clinical populations including congestive heart failure, stroke, and asthma patients. We are working on analyzing the data using a variety of methods to uncover promising digital biomarkers. Real-time machine learning models are being used to identify and classify the signals.

We are also developing complementary sensors for sweat and saliva testing to provide a more holistic approach to patient monitoring. We are working towards syncing all the data together to provide comprehensive and real time information of the patient's well-being, to alert of (and eventually hopefully prevent) potential deterioration.

598928 (Invited Plenary Talk): Self-Healable Ultra-Stretchable Wearable Sensor Materials

Abstract:

Wearable strain sensors are essential for the realization of applications in the broad fields of remote healthcare monitoring, soft robots, immersive gaming, among many others. These flexible sensors should be comfortably adhered to skin and capable of monitoring human motions with high accuracy, as well as exhibiting excellent durability. However, it is challenging to develop electronic materials that possess the properties of skin. The presented skin-like electronic material exhibits ultrahigh stretchability, repeatable autonomous self-healing ability, quadratic response to strain, and linear response to flexion bending. This conductive polymer system, under ambient conditions, synergistically constructs a regenerative dynamic polymer complex crosslinked by hydrogen bonds and electrostatic interactions, which enables these unique properties. Sensitive strain-responsive mechanisms owing to the homogenous and viscoelastic nature provide omnidirectional tensile strain and bending deformations. Furthermore, this material is scalable and simple to process in an environmentally-friendly manner, paving the way for the next generation wearable sensors.

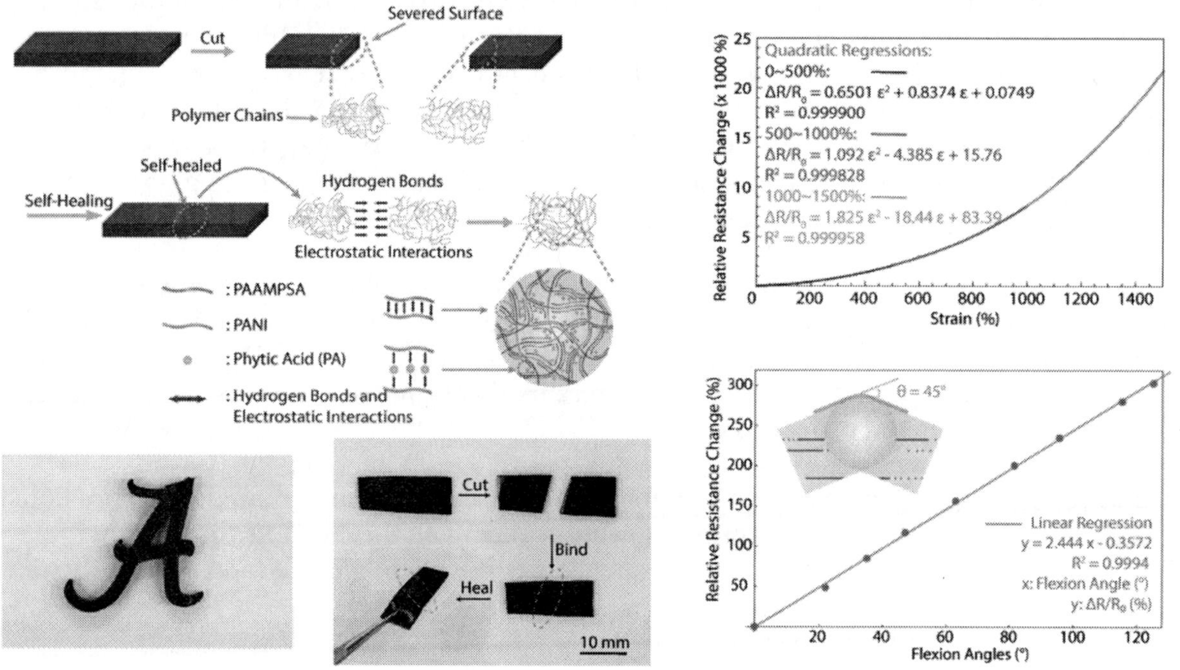

AUTHOR INDEX